WORLD
OCEAN CENSUS

WORLD OCEAN CENSUS

A GLOBAL SURVEY OF MARINE LIFE

DARLENE TREW CRIST * GAIL SCOWCROFT * JAMES M. HARDING, JR.

CENSUS OF MARINE LIFE

FIREFLY BOOKS

A FIREFLY BOOK

Published by Firefly Books Ltd. 2009

First printing

Publisher Cataloging-in-Publication Data (U.S.)
Crist, Darlene Trew.
World ocean census : a global survey of marine life / Darlene Trew Crist, Gail Scowcroft, James M. Harding, Jr.; foreword by Sylvia Earle.
[256] p. : ill., photos. (chiefly col.) ; cm.
Includes bibliographical references and index.
Summary: Focuses on the Census of Marine Life, a 10-year global project to investigate our oceans and the creatures that live there. Examines how the Census was carried out and what the researchers discovered.
ISBN-13: 978-1-55407-434-1   ISBN-10: 1-55407-434-7
1. Underwater exploration. 2. Marine biology. 3. Oceanography.   I. Census of Marine Life. II. Scowcroft, Gail. III. Harding, James M., Jr.
IV. Earle, Sylvia A., 1935- . V. Title.
574.9202 dc22   QH91.17C757   2009

Library and Archives Canada Cataloguing in Publication
Crist, Darlene Trew
World ocean census : a global survey of marine life / Darlene Trew Crist, Gail Scowcroft and James M. Harding.
Includes bibliographical references and index.
ISBN-13: 978-1-55407-434-1   ISBN-10: 1-55407-434-7
1. Census of Marine Life (Program). 2. Marine animals — Counting — Methodology. 3. Marine biology — Research — Methodology.
4. Zoological surveys — Methodology. 5. Marine ecology. 6. Oceanography. I. Scowcroft, Gail II. Harding, James M., 1975- III. Title.
QH95.C74 2009   578.77   C2009-900901-3

Published in the United States by
Firefly Books (U.S.) Inc.
P.O. Box 1338, Ellicott Station
Buffalo, New York 14205

Published in Canada by
Firefly Books Ltd.
66 Leek Crescent
Richmond Hill, Ontario L4B 1H1

Cover and interior design by Sari Naworynski
Photo Editor: Nicole Caldarelli

Page 2: Angels in a dark sea, Cliona limacina. *These angelic-looking swimming snails, or pteropods, were collected in the Canada Basin. Pteropod means "wing-footed," which refers to the modification of the snail's foot that allows it to swim through the water.*

Printed in China

The publisher gratefully acknowledges the financial support for our publishing program by the Government of Canada through the Book Publishing Industry Development Program.

FSC

Mixed Sources
Product group from well-managed
forests, controlled sources and
recycled wood or fibre

Cert no. DNV-COC-000054
www.fsc.org
© 1996 Forest Stewardship Council

THIS BOOK IS DEDICATED TO RANSOM A. MYERS

AND ROBIN RIGBY, OUR LATE COLLEAGUES.

THEIR CONTRIBUTIONS WERE MANY,

THEIR WORK WAS INSPIRATIONAL

AND THEIR INSIGHTS ARE SORELY MISSED.

ACKNOWLEDGMENTS

Our appreciation is extended to Lionel Koffler, president of Firefly Books, for the original concept of this book. We are also grateful for his continued support and encouragement.

This book would not have been possible without the gracious co-operation of our Census of Marine Life colleagues around the world. We are beholden to the many researchers who carefully and diligently worked with us to ensure that their findings were accurately portrayed and to the many photographers who so generously shared their wonderful images.

We are indebted to Ron O'Dor, co-senior scientist of the Census of Marine Life, who added to his already hectic schedule by reviewing and editing our manuscript. Our sincere thanks are also extended to Sara Hickox, director of the University of Rhode Island's Office of Marine Programs, for her support and encouragement. Without her endorsement, this book would not have been possible. We also must recognize the many and significant contributions of Jesse Ausubel of the Alfred P. Sloan Foundation, whose vision helped create the Census of Marine Life and who continues to serve as a source of inspiration for all who have the privilege of working closely with him.

We are honored to tell the story of the Census of Marine Life, for it is an ambitious, far-reaching and innovative undertaking that will significantly contribute to the understanding of the global ocean and the creatures that live within it.

Opposite: Ptychogastria polaris. *This beautiful jellyfish, usually a deep-dwelling species, can be seen near the surface in the Arctic and Antarctic.*

The appropriately named vampire squid (Vampyroteuthis infernalis; *literally,
"vampire squid from hell"*) defends itself by spitting a sticky cloud of bioluminescent mucus from the ends of its arms, presumably dazing would-be predators and
allowing its escape into the darkness.

Contents

FOREWORD

BY SYLVIA EARLE, PhD

AMBASSADOR FOR THE WORLD'S OCEANS AND NATIONAL
GEOGRAPHIC SOCIETY EXPLORER-IN-RESIDENCE

The only other place comparable to these marvelous nether regions,
must surely be naked space itself ... where the blackness of space,
the shining planets, suns and stars must really be closely akin to
the world of life as it appears to the eyes of an awed human being,
in the open sea half a mile down.
- WILLIAM BEEBE, HALF MILE DOWN

Until recently, many had the impression that the diversity of life in the sea
was much lower than the complexity and richness of life on the land, especially tropical rainforests and temperate woodlands. It is easy to understand
how this misconception arose – largely because of the immense diversity
of insects, especially beetles, altogether comprising about half of all the
species thus far named that live on land. More important, humans are terrestrial, air-breathing creatures able to travel everywhere over the face of
the land, even to the highest mountains, driest deserts and coldest polar
regions, and thus are familiar with the various creatures great and small
that live on land. Accessing the sea is more challenging.

Although the ocean comprises 99 percent of the biosphere and only
5 percent of that has been seen – let alone explored – it is little wonder
that misconceptions have developed about the true nature of life on
Earth. Fortunately, new technologies enhance our ability to find and
document life from sunlit reefs to the deepest, darkest seas – and the
results are magnificently portrayed in this volume. Great challenges
remain, but those involved with the Census of Marine Life have made

giant strides in evaluating the nature of the past, present and future of life in the sea, and thus of life on Earth.

Far from being monotonous empty space, as some have long believed, the ocean everywhere is alive. In a single swirl of saltwater, a plankton-feeding whale shark may swallow the larval or adult stages of 15 or more phyla (or divisions) of animals – as many as all terrestrial phyla of animals combined. That single gulp also will likely contain a dozen or so phyla of Protists, including minute photosynthetic forms that do much of the heavy lifting in terms of generating oxygen and transforming water combined with carbon dioxide into food. Then there are the microbes. Thousands of new species are being discovered in nearly every sample of seawater being analyzed with current techniques, from bacteria to members of the kingdom Archeae, discovered late in the 20th century. It is not just the small creatures that have gone unnoticed. New families as well as new genera and species of coral, sponges, echinoderms, annelids and others are turning up on nearly every dive below a thousand feet or so. When biologist-diver-explorer Richard Pyle ventures into the almost-light almost-dark region of the sea known as "the twilight zone," he finds new fish species at the rate of about seven per hour of observation.

Given the magnitude of new discoveries, and the vast areas of ocean yet to be explored, it is clear that the unknown species of marine life far exceed the 250,000 or so now accounted for. So, how many kinds of plants, animals, microbes and other forms of life are in the sea? Estimates range from a million to one hundred million, making the goal of the Census of Marine Life – taking stock of the diversity of life in the sea – one of the most ambitious undertakings in the history of humankind. In effect, it means exploring, analyzing and making some kind of systematic sense of most of life on Earth. It is a daunting but worthy goal, vital, in fact, if humankind is to achieve an understanding of the natural systems that make life on Earth possible.

The importance of the Census is made urgent because at the same time that more is being learned about the diversity of life in the sea than during all preceding history, more is being lost. Jacques Cousteau's personal perspective – from pioneering dives in the 1950s in pristine seas to an era a few decades later of "paradise lost" – inspired the world to take notice, and to act. It is not just the obvious decline of marine mammals, seabirds, fish and other wildlife owing to deliberate or incidental killing of ocean wildlife for food and products, although some, such as the Caribbean monk seal, last seen in 1952, have been eliminated thus. Numerous creatures have a narrow home range, and when their habitat is destroyed, they are lost as well. Species endemic to specific coral reefs, to individual seamounts and to highly specialized associations, such as barnacles that live only on one kind of turtle or whale or crab, are especially vulnerable. Changes in ocean temperature and chemistry through human actions, now including acidification, are causing comprehensive shifts of geological magnitude on the global environment, and in turn on the species that comprise the countless bits and pieces that make up the whole.

No one can be certain of what the consequences will be of the damage that humans have wrought to the underpinnings of what has made Earth a hospitable place for our species. No one will ever know how many kinds of creatures have been or will be destroyed as a consequence of these changes. But it is clear that maintaining the diversity of life, from individual species to large ecosystems, is the key to resilience, to holding the planet steady through the trauma of rapid climate shifts and the other unprecedented changes that are upon us. At present, about 12 percent of the land is invested in national parks and preserves, worldwide, protecting the diversity of terrestrial and freshwater species and systems while but a fraction of 1 percent of the ocean is similarly safeguarded.

In an exceptional model of international cooperation targeting a singular goal, more than two thousand Census of Marine Life scientists have delved into past records, dived into modern oceans, and extrapolated future scenarios. Together, they have yielded a priceless gift: enhanced understanding of the nature of life in the sea and its importance concerning all that humans care about – health, the economy, security and, most important, a planet that works in our favor. A distillation of their Herculean decade of exploration is contained here, magnificently illustrated and eloquently written by the authors and contributors.

Some will treasure *World Ocean Census* as a valuable reference, others as a place to find white-knuckle adventures. The images alone will cause many to re-evaluate their concepts of what astonishing forms are embraced within the bounds of what constitutes an eye, a heart, a body of living tissue. The underlying similarities shared by all living things – humans very much included – shine through, while maintaining wonder at the infinite capacity for diversity: from the broad divisions of life to the individual speckles and shapes that distinguish each sardine, salp and starfish from every other of its kind. Above all, the breakthroughs in knowledge gained, and awareness of the magnitude of what remains to be discovered, inspire hope that the greatest era of ocean exploration – and ocean care – will now begin.

Unlocking the Mystery of What Lives Below the Surface of the Global Sea

For thousands of years, indigenous traditions have described the ocean as the mythological birthplace of life on Earth, a mysterious place filled with life-sustaining forces. In an ancient myth of the Yurok Indians of California, two great beings, Thunder and Earthquake, worked together to create the ocean and fill it with water. The animals came to live there because of its beauty. Their story speaks of seals that came to the newly formed ocean "as if they were thrown in by handfuls." Looking upon the ocean basin they had made – vast, deep and full of water – Earthquake and Thunder were satisfied that their job was done. The ocean was large enough to provide sustenance for all of Earth's creatures.

It is not surprising that the ocean has long fascinated humankind. Earth is the only planet known to have liquid water at the surface, and very little of the ocean has been scientifically investigated. However, there are accepted theories about its formation and composition and even some consensus about how it works, despite the lack of comprehensive data. Most scientists believe that the first, shallow oceans formed between 4 billion and

Opposite: Earth as it appears from a satellite orbiting high above the Indian Ocean.

3.5 billion years ago. As the molten-hot, newly forming crust of Earth cooled, it gave off great volumes of steam and water vapor, which in turn caused clouds to form and rain to fall. The rain carried salts and other elements from Earth's cooling surface into shallow depressions or basins in the crust. Before an ocean could form, the temperature had to fall below 100°C, the boiling point of water, so that the liquid could remain stable.

Over time and through complex geological processes, cooled crustal plates formed over the more molten interior mantle, and the ocean basins deepened. As the plates slowly circulated over the molten mantle, they drew together and were pulled apart over and over again, eventually forming Earth's first supercontinent – a single landmass consisting of all the modern continents – Vaalbara. Believed to have formed between 3.6 billion and 3.3 billion years ago, it was surrounded by a vast sea. As this early supercontinent broke apart and the crustal plates continued their journey, the configuration of the ocean changed as well. As we know it today, the world's ocean accounts for roughly 71 percent of our planet's surface and 99 percent of its inhabitable volume; its average depth is approximately 3.8 kilometers (close to 2.4 miles).

Scientists are still trying to unravel one of the greatest mysteries of Earth: when did "life" first appear and how did it happen? The scientific community has been debating the timing and mechanisms of the origin of life for many years. Most agree that the earliest life-forms evolved in the ocean, most likely as primitive, one-celled forms that appeared about 3 billion years ago. This primitive life was all that existed for approximately the next 2 billion years. Then a profusion of multicellular life exploded and began to fill the oceans. As some marine species became able to live on land, new and increasingly complex forms of life began to appear all over the planet.

Many of the microscopic life-forms, or microbes, that currently live in the ocean may be similar to Earth's early life, and their abundance is extraordinary. Census microbiologist Mitch Sogin and his team at the Marine Biological Laboratory at Woods Hole, Massachusetts, discovered more than 20,000 kinds of microbes in a single liter (about 1 quart) of seawater; previous research had led them to expect only 1,000 to 3,000. "Peering through a laboratory microscope into a drop of seawater is like looking at the stars on a clear night," says Census of Marine Life marine microbiologist Victor Gallardo, of Universidad de Concepción, Chile. "New DNA tag-sequencing technology increases resolution much like the Hubble Telescope did for space. We can see marine microbial diversity

Opposite: Census of Marine Life scientists are investigating marine life from microbes to whales, from top to bottom, from pole to pole. Shown here is a white-topped coral crab, collected from dead coral off Heron Island, Australia, on the Great Barrier Reef.

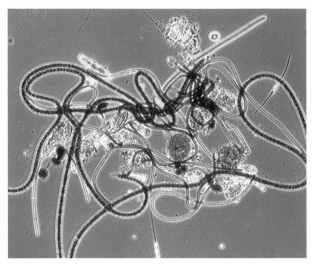

Marine microbes are the oldest life-forms. In this collection the darkest ones are a very common filamentous form that has not yet been identified in the scientific literature. The large pink oval is a cell of Chromatium, *a purple sulfur bacterium; the green is a cyanobacterium, one of a group of blue-green bacteria capable of photosynthesis. The curving structure at about two o'clock is a diatom* (Nitzschia *species), a microscopic algal species that is a photosynthetic eukaryote – a single-celled or multi-cellular organism whose cells contain a distinct nucleus.*

to which we were blind before. These rare ancient organisms are likely to prove a key part of nature's history and strategy."

The current average salt content of this cradle of life is approximately 35 parts to 1,000 parts of ocean water. Some scientists estimate that the oceans contain as much as 50 quadrillion (50 million billion) tons of dissolved solids, mainly dissolved salts such as sodium chloride. If the salt in the sea were removed and spread evenly over the earth's land surface, it would form a layer more than 150 meters (500 feet) thick, about the height of a 40-story office building! Salts become concentrated in the sea because the sun's heat evaporates almost pure water from the ocean surface, leaving the salts and salty water behind. This process is part of a continual exchange of water between the earth and the atmosphere called the hydrologic cycle.

The salinity of ocean water varies. It is affected by such factors as melting ice, inflow of river water, evaporation, rain, snowfall, winds, wave motion and ocean currents that cause horizontal and vertical mixing of the saltwater. Small changes in salinity can greatly affect ocean life. Some organisms can tolerate wide swings in salinity, while others, such as many corals, can tolerate only narrow salinity ranges. The saltiest water occurs in the Red Sea and the Persian Gulf, where rates of evaporation are very high. Of the major ocean basins, the North Atlantic is the saltiest. Low salinities occur in polar seas, where the salt water is diluted by melting ice and frequent precipitation.

Salinity is not the only variable that affects marine life. Other parameters include temperature, available nutrients, currents, winds, storms and ice. To fully comprehend how species survive in the ocean, it is imperative to understand how the ocean works. Important unanswered questions remain about the interdependency of oceanic parameters, such as how winds affect currents, how currents affect the temperature and flow of nutrients, and how nutrients control productivity. The limits to our current understanding can be overcome only by further ocean exploration – by initiatives such as the Census of Marine Life that are investigating the biology of the global ocean and how its physical properties affect its inhabitants.

Launched in 2000 with the goal of producing the first-ever ocean census by 2010, the Census of Marine Life brought together 2,000 explorers from 82 nations to answer three important questions: What

Lace corals, such as this Stylaster *species from Palau, Micronesia, are vulnerable to changes in salinity. The organisms that depend on them, such as this tiny unidentified reef crab, approximately 2.5 centimeters (1 inch) across, are thus also at risk.*

once lived in the global ocean? What is living there now? What will live there in the future? So little is known about what lives below the surface in the global ocean that the Census is much like a space program; it too has inherent challenges and dangers, and likewise generates excitement about exploring the unknown. To reach the depths of the global ocean – as deep as 11 kilometers (7 miles) below the surface – requires equipment that is just as sophisticated as the technology used to explore the outer reaches of the universe. Unlike space programs, however, underwater technology has only recently become advanced enough to allow explorers a fish's-eye view of what lives in the deepest, darkest, coldest reaches of the ocean.

Census of Marine Life scientists have been working at the cutting edge, using new technology to explore previously unreachable places and capturing images of incredible life-forms – everything from blind lobsters to worms that live without oxygen – that exist at the bottom of the sea. This book provides an inside look at their many adventures, reveals their findings, and shares the fabulous images they have captured as they unlock the mystery of what lives below the surface in the wondrous global sea.

PART ONE

WHAT LIVED IN THE OCEAN?

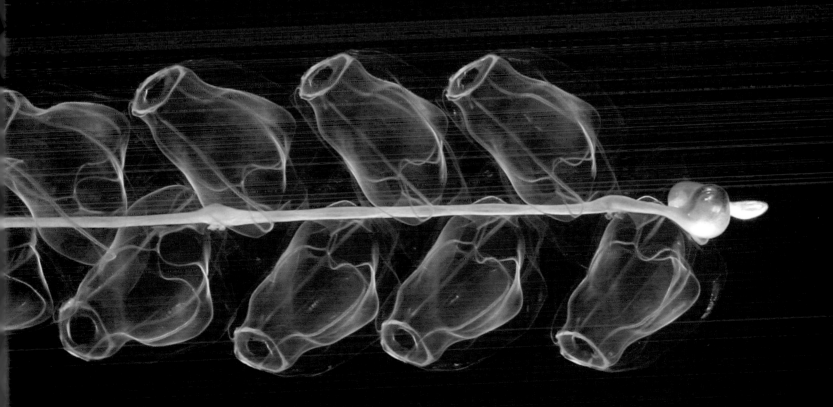

CHAPTER ONE

The Known, the Unknown and the Unknowable

The dream of knowing what lives in the sea is ancient, vast, and romantic.
What is new is the urgency of the task, the ability to carry it out,
and the fact that growing numbers of us are trying.
– Jesse H. Ausubel, Program Director, Alfred P. Sloan Foundation

In 1997, about 20 of the world's leading marine ichthyologists – scientists who study fish – sat down together around a table at the Scripps Institution of Oceanography in La Jolla, California. Their task was to assess what was known and what was unknown about the diversity of marine fishes. As their discussions unfolded, it didn't take them long to conclude that much more research was needed to determine the unknown. In fact, so little is known about what lives in the seas, scientists often joke that more is known about the surface of other planets than about the bottom of Earth's ocean.

These 20 men and women decided to shift their focus from studying known species to uncovering those that were unknown, and perhaps even beginning to define what may *never* be known about what lives below the surface of the sea. They began to lay the groundwork for a baseline of knowledge about marine life in the global ocean from which changes could then be measured. An overall survey of marine species from the ocean's surface to the bottom, including what lives in the sediments of the seafloor, had never before been undertaken. Many of the

Pages 20–21: A physonect siphonophore, Marrus orthocanna, *photographed during the National Oceanic and Atmospheric Administration's Arctic "Hidden Ocean" expedition in support of the Census of Marine Life.*

Opposite: Census of Marine Life scientists are often privy to the unseen wonder and beauty of what lives below the surface, as exemplified here by a school of common bluestripe snapper (Lutjanus kasmira) *and shoulderbar soldierfish* (Myripristis kuntee) *swimming in this Hawaiian reef.*

This fabulously marked polychaete, or marine tube worm, Loima *sp.,*
inhabits the waters off Lizard Island, Queensland, Australia.

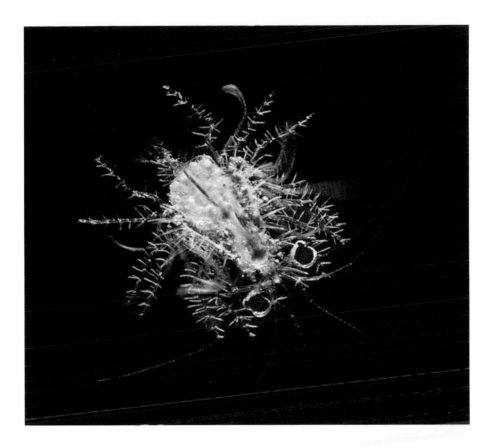

The spines on this crab larvae (spiny decapod megalops), although beautiful, serve as protection and camouflage.

scientists in the room that day viewed this change in focus as an opportunity to enter an exciting new age of discovery, comparable to those of Darwin, Linnaeus and even the courageous Captain Cook.

The initial La Jolla group of 20 was lucky to have among them a potential benefactor for whom the concept of a census of marine fishes was not a new idea. About a year before, J. Frederick Grassle, head of the Institute of Marine and Coastal Sciences at Rutgers University, had approached Jesse H. Ausubel, a program director for the Alfred P. Sloan Foundation, in his summer office at the Marine Biological Laboratory in Woods Hole, Massachusetts. Grassle pleaded the case for learning about the vast unknowns of marine life, a case he made effectively. His scheduled hour-long meeting turned into an afternoon's worth of discussion in which the seeds of a census of marine life were sowed.

The meeting at Scripps reinforced the need for such a census, and Ausubel took up the gauntlet. He urged the Sloan Foundation board to support the concept of a census of marine life, emphasizing the chances for exciting discoveries, the importance of establishing baseline information on the distribution of marine biodiversity, and the fact that the changing abundance of many species pointed to a critical

need for improved management of fisheries and marine resources, which the census theoretically could help accomplish. The board was convinced.

With the Sloan Foundation's backing, scientists from literally all over the world were invited to join in the discussions about how to tackle the daunting challenge of censusing the global oceans. From the outset it was clear that such an undertaking would be an enormous one, requiring not only huge amounts of resources, capital and intellect, but also the vision and cooperation of all those who would choose to participate. It was an idea that sparked the imagination of many marine scientists around the globe.

THE GREAT UNKNOWN

Even now, at the beginning of the 21st century, 95 percent of the world's ocean basins and seas has yet to be explored (some put this figure as high as 98 percent). Part of the reason is simply the global ocean's vast size: it comprises approximately 71 percent of the planet's surface and covers 361 million square kilometers (139 million square miles). And there is more to the world's ocean than meets the eye – a vast story unfolds below the surface. The global volume of ocean water is 1,370 million cubic kilometers (329 million cubic miles), with an average depth of 3.8 kilometers (2.4 miles). The deepest ocean trench areas extend 10.5 kilometers (more than 6.7 miles) below the sea surface. And if the obstacles of size, volume and mass were not enough, other deterrents to exploration – darkness and pressure – greatly increase the challenge, cost and risk for those who dare to venture below the surface. Only recently have technological advances allowed scientists to successfully tackle the physical challenges of exploring dark ocean extremes at intense pressures.

Studying the world's ocean is further compounded by the fact that all oceans are really one vast body of water. Each of the five ocean basins – the Pacific, Atlantic, Southern, Indian and Arctic oceans – are interconnected by major surface and deep-water currents in a circulation system that creates a single body of water. All ocean life is connected by this system, so we have to understand it in order to understand its biodiversity (see "The Global Ocean Conveyer Belt" on page 31).

Estimates of the number of marine species in the global ocean are uncertain at best, ranging from one to ten million. Even when the search is

Opposite: This spectacular blue-eyed hermit crab (Paragiopagurus diogenes) *is an example of Census discoveries that raise more questions than answers. The shiny gold on the claws of this crab, captured in the French Frigate Shoals off the Northwestern Hawaiian Islands, is a phenomenon not seen before. Scientists believe it serves as a form of communication. Attached to its shell, the crab also has its very own species of anemone (the fuzzy brown area underneath), which is not known to attach to any other species of hermit crab.*

limited to fish, the number of total marine species can't be determined with certainty. About 15,000 species of marine fishes have been identified, and ichthyologists estimate that another 5,000 or so fish species remain to be discovered. When the critters get smaller, the level of uncertainty about their actual numbers increases proportionately. Less than 1 percent of the microscopic marine life that exists in all global ocean water, for example, has been identified. There are also many unknowns about life in specific ocean habitats, even though the habitats may have been described. In coral reefs, for example, scientists estimate that less than 10 percent of reef life has yet to be identified. Because of coral bleaching and other threats to reef habitats, some species may be lost before ever being found.

This lack of knowledge about what inhabits the oceans, where species live, and in what numbers they are present creates a huge challenge for managing fisheries and other ocean resources. Information on the status of known fish and shellfish species and population trends is limited to the 200 or so commercially important species, including tuna, salmon, scallops and a few species of whales, for which the most data are available. Abundance estimates are derived mostly from catch statistics, which are usually reported by the fishers themselves. So, even if the number of marine species is closer to 1 million than the predicted 10 million, knowing about the diversity (number of species), distribution (where they live) and abundance (how many individuals) of only 200 marine species is simply not adequate from either a scientific or a management point of view.

THE CENSUS OF MARINE LIFE

Three years after the idea of a global ocean census was first conceived, 60 researchers from 60 different institutions in 15 countries had joined a fledgling effort – the Census of Marine Life – to "assess and explain the diversity, distribution, and abundance of marine life." Their objectives were grand and included investigating everything that lived in the global oceans, from microbes to whales, from top to bottom, from pole to pole. Ultimately the goal was to determine how marine animal populations have changed and will change over time. (By 2008 the number of participants had grown to 2,000 scientists from 81 nations, and the financial commitment toward the project had topped $500 million.)

To put these broad goals into context, it helps to understand what the Census hopes to accomplish in each category:

Diversity: The Census intends to compile the first comprehensive list of all forms of life in the global ocean – the first of its kind. Its second diversity objective is to estimate how many species will remain unknown, that is, yet to be discovered.

Distribution: The Census hopes to produce maps displaying where marine animals have been observed, with their ranges or territories – where they could possibly live. The latter is particularly significant in terms of where animals might be able to live if climate changes continue to occur.

Abundance: The goal is to estimate many populations of marine life in numbers or by weight – also known as biomass.

To assess the diversity, distribution and abundance of marine life, the Census of Marine Life was structured around three questions: What has lived in the global ocean? What lives in the global ocean? What will live in the global ocean?

This deepwater jellyfish (Crossota norvegica) *was photographed during a Census of Marine Life expedition to the deep Canada Basin in 2005.*

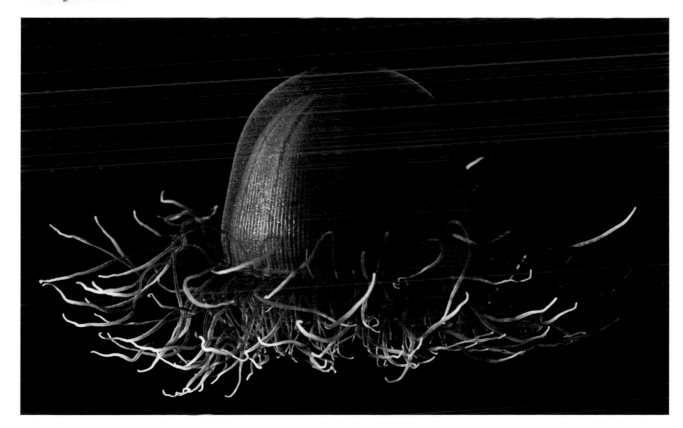

Cold, salty, dense water (the light blue line) sinks in Earth's northern polar region and heads south along the western Atlantic basin..

The current is "recharged" as it travels along the coast of Antarctica and picks up more dense, cold, salty water.

THE GLOBAL OCEAN CONVEYOR BELT

The sea is, in fact, one ocean — one ocean with five great names and a thousand little ones. There is no real boundary to any sea, save continental land. The waters mingle everywhere and the names are geographic for convenience only. — ALAN VILLIERS, *OCEAN OF THE WORLD* (1963)

The global ocean is truly one great body of water composed of five oceans — the Pacific, Atlantic, Southern, Indian and Arctic — and several smaller seas. All of the ocean basins and seas are interconnected and together cover approximately 71 percent of Earth's surface. This great global ocean contains approximately 97 percent of the water on Earth.

The complex circulation of the global ocean's salty waters is controlled by winds that drive surface water and by the cooling and sinking of waters in the polar regions to form deepwater currents. A drop of water carried into the North Atlantic by warm surface-water currents will eventually cool and sink. This drop then gets caught in deepwater currents and becomes entrained in the great ocean "conveyor belt" system, where it is carried along

for thousands of miles. Eventually it will surface again back in the North Atlantic — a journey that takes approximately a thousand years to complete.

Winds drive ocean currents in the upper 100 meters (330 feet) of the ocean's surface, but currents also flow thousands of meters below the surface. These deep-ocean currents are driven by differences in the water's density, which is controlled by temperature and salinity in a process known as thermohaline circulation.

The global ocean circulation system is powered by many different physical factors. Surface circulation carries warm tropical surface waters toward the poles. Heat is dispersed from these waters along the way and absorbed by the atmosphere. As ocean surface water approaches polar regions

The main current splits into two sections, one traveling northward into the Indian Ocean while the other heads up into the western Pacific.

The two branches of the current warm and rise (the red lines) as they travel northward, then loop back around southward and westward.

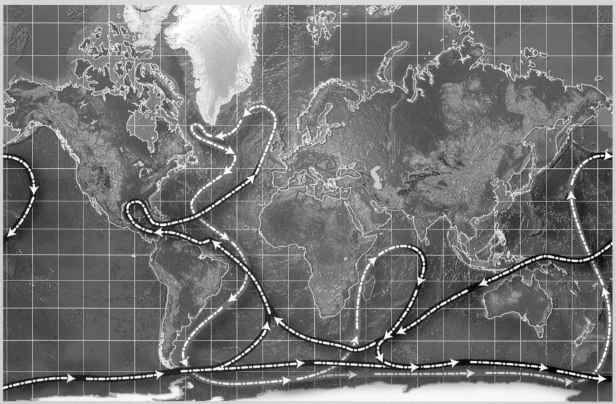

The now-warmed surface waters continue circulating around the globe, eventually returning to the North Atlantic, where the cycle begins again.

during the winter months, it is further cooled and sinks to join deepwater currents, especially in the North Atlantic and near Antarctica.

Cold deep-ocean water gradually warms up and becomes less dense as it travels toward the equator, eventually rising again to the ocean surface. Once there, it is carried back toward the polar regions in surface current, where it cools and sinks and the cycle begins again. The pace of this cycle determines the amount of time during which heat is transferred from the warm ocean waters to the atmosphere. If the pace of the global ocean conveyor belt slows down, there will be more exchange of heat between the ocean and the atmosphere, contributing to further warming of Earth's climate.

In Earth's polar regions, ocean water freezes into sea ice. The surrounding seawater gets saltier because when sea ice forms, the salt is left behind. As the seawater gets saltier, its density increases and it starts to sink. Surface water flows in to replace the sinking water and, in turn, it also eventually becomes cold and salty enough to sink. This initiates the deep-ocean currents that drive the global ocean conveyer belt.

This deep water moves south between the continents, past the Equator and down to the tips of Africa and South America. As the current travels around the edge of Antarctica, the water cools and sinks again, as it does in the North Atlantic. Thus the conveyor belt gets "recharged."

As it moves around Antarctica, two sections split off the "conveyor belt" and turn northward. One section moves into the Indian Ocean and the other into the Pacific Ocean. These two sections warm up and become less dense as they travel northward toward the Equator, so they rise to the surface. They then loop back southward and west-ward to the South Atlantic and eventually return to the North Atlantic, where the cycle begins again.

The global conveyor belt is a vital component of the ocean's nutrient and carbon dioxide cycles. Warm surface waters become depleted of nutrients and carbon dioxide, but they are enriched again as they travel via the conveyor belt. The world's food chain depends on the cool, nutrient-rich, upwelled waters that support the growth of algae and seaweed.

Variations in the ocean's circulation can affect weather patterns. One important variation is a change in the eastern Pacific equatorial region. Strong El Niño events here can dramatically affect marine life. These events occur when a cap of warm water arriving on an easterly-flowing equatorial current prevents cool, nutrient-rich waters that fuel the food chain from rising to the surface along the coast of Peru. Large populations of fish crash, leaving no food for seabirds and marine mammals – just one example of the interconnectedness of ocean circulation and the health of marine animal populations throughout the global ocean.

Research suggests that the conveyor belt may be affected by climate change. If global warming results in increased rainfall in the North Atlantic, along with the melting of glaciers and sea ice, the influx of warmer, fresh water on the sea surface could slow down the formation of sea ice, disrupting the pace of the sinking of cold, salty water. These events could result in further climatic changes.

Above: Global climate change could disrupt the global conveyor belt, causing potentially drastic temperature changes in Europe and even worldwide, disrupting and threatening marine life in the global ocean.

WHAT LIVED IN THE OCEAN?

The first step in completing a census of marine life is to look at the past. Census scientists took up the challenge of constructing the history of marine animal populations since human predation became important – roughly over the past 500 years. Poring over dusty old records from monasteries, ships' logs and even restaurant menus, teams of fisheries scientists, historians, economists and others searched for data that had previously been hidden from view. They launched case studies of specific species in southern Africa, Australia and about a dozen other regions around the globe. The idea was that, when pulled together, these case studies would paint the first reliable picture of life in the oceans "before fishing." Such long historical records of marine populations could help distinguish between the contributions of natural fluctuations in the environment and the effects of human activities, an understanding that could prove critical in setting goals for the protection of marine species in the future.

WHAT LIVES IN THE OCEAN?

Determining what lives in the global ocean is a complex undertaking. To tackle it, the Census established 14 field projects: 11 cover major habitats such as ice oceans and the ocean bottom, while 3 are globally focused on species groups. Everything from microbes to whales was included in the investigation.

To study habitats and species at this global level, several technologies were employed, including everything from acoustics (sound) to

What lived in the ocean? In 1946, hundreds of tuna were offered for sale in a fish auction hall in Skagen, Denmark, well before a collapse in the bluefin tuna population in the 1960s, which has been documented by Census researchers.

optics (cameras), tagging and genetics, supported by physical sampling. The different technologies allowed researchers to see and experience the ocean from an animal's point of view, and each brought a different perspective to the experience.

Sound helped researchers survey large ocean areas. Light captured details that otherwise would have been lost in the dark. Photographs and video provided information when depth, pressure and other elements made it impossible to capture a live specimen. Genetics provided a means to identify species when whole specimens were not available or morphological comparison was inadequate. Sampling provided scientists with real specimens to study in order to confirm identification and learn more about what makes a specific animal tick. Together these technologies acted as a toolbox from which an animal's view of the ocean could be constructed and better understood.

Using high-powered microscopes and advanced digital cameras, Census scientists are able to see what was once invisible, or nearly so. This photograph shows a microscopic zooplankton.

Hydrothermal vents – places on the ocean floor where hot, chemically enriched water flows up through cracks in the Earth's crust – support rich and diverse forms of marine life that are fueled by chemical energy rather than solar energy. Zoarcid fish, or eelpouts, swim over a community of the tubeworm Rifitia pachyptila *on an East Pacific Rise hydrothermal vent. The crabs are* Bythograea *species.*

Among hundreds of animals collected in a little-explored area of the deep Arctic Canada Basin was this as yet unidentified anemone, collected by a remotely operated vehicle at a depth of more than 2,000 meters (6,600 feet). The specimen is about 5 centimeters (2 inches) high and 3 centimeters (1 inch) in diameter.

Once a framework was established for studying the global ocean, Census scientists proceeded with yet another large challenge: figuring out how to work together across different time zones, continents and cultures. It was a human experiment that produced surprisingly good results. English is used as the common language of the Census. The Internet, videoconferencing and email became staples of communication. Each expedition was a floating United Nations, with an international group of scientists sharing their knowledge, experience, living space and cultural idiosyncrasies. Because of virtually seamless international communication systems, even when expeditions were in the middle of the ocean well away from land, satellite communications enabled Census scientists to share data not only among themselves but

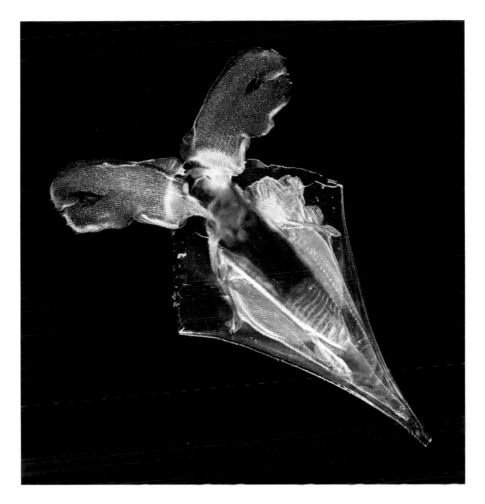

*This zooplankton species (*Clio pyramidata, *a pteropod, or swimming snail) was in the first group of animals ever to have their genes sequenced at sea. This feat was accomplished by a team of international Census scientists in the middle of the Atlantic Ocean.*

also, by going online, with others in the scientific community and the general public.

The international aspect of the Census has made for some interesting configurations. For example, during a 10-week expedition in 2006 aboard the German research vessel *Polarstern*, 52 marine explorers from 14 countries completed the first comprehensive biological survey of a 10,000-square-kilometer portion of the Antarctic seabed. Data were transmitted to the Australian Antarctic Division in Tasmania, Australia, and communications were coordinated through the Alfred Wegner Institute for Polar and Marine Research in Bremerhaven, Germany.

Another team of 28 Census experts from 14 nations trawled rarely explored tropical ocean depths between the southeast U.S. coast and the Mid-Atlantic Ridge to inventory and photograph the variety and abundance of zooplankton. Of the thousands of captured specimens, 220 were DNA-sequenced at sea, revealing a number of new species. The team members, who had each spent decades learning to distinguish species within a particular group, sorted through samples in a kind of an international assembly line "that would have made Henry Ford proud," according to University of Connecticut postdoctoral investigator Rob Jennings, leader of the onboard "Team DNA."

What Will Live in the Ocean?

The answers to the broader and perhaps more complex question of what will live in the world's ocean in the future will depend upon what is found in historic and present investigations, and will require sophisticated modeling and simulation techniques to help determine the future. The Census project devoted to predicting the future of marine animal populations has created new statistical and analytical tools that make it possible to integrate data from many different sources around the globe. Thus far the conclusions have been sobering.

One study projected the end of the global commercial fishery by 2050 if current trends persist. Another reported that the diversity of major predatory fish in the open ocean has declined because of overfishing by as much as 50 percent in just 50 years. Another addressed how changes in the large predator populations were creating changes in populations of the smaller species that form the foundation of the food web. Studies such as these and others have reinforced the need for securing scientifically based information to better manage fisheries, conserve species diversity, reverse habitat loss and reduce the impacts of pollution in the world's ocean, and possibly to respond to global climate change in intelligent, informed ways.

From the project's outset, the Census founders wanted to share with a broad audience what was being learned. They chose to make the data public as it unfolded, via an interactive website that serves as a repository for all Census data. An Internet database called OBIS (Ocean Biogeographic Information System) contains more than 16 million records of marine life and is growing every day. Projections are that OBIS may contain upward of 20 million records or more when the first Census of Marine Life is completed.

Certainly much has been accomplished since the idea was conceived in 1997, and much more remains to be discovered. Perhaps the most exciting element is that the idea of learning more about what lives below the sea surface has captured the imagination of scientists and

citizens from around the world. The prospect of discovery and of contributing to predicting the future of marine life has intrigued scientists, administrators and funders enough to create this extraordinary international collaboration. Census scientists carry the vision of determining the known and unknown – and of recognizing the unknowable – about what lives in the global ocean. Many new discoveries are yet to come.

This flamingo tongue snail (Cyphoma gibbosum) *was photographed near Grand Cayman, British West Indies. It is listed in the Gulf of Mexico biodiversity inventory.*

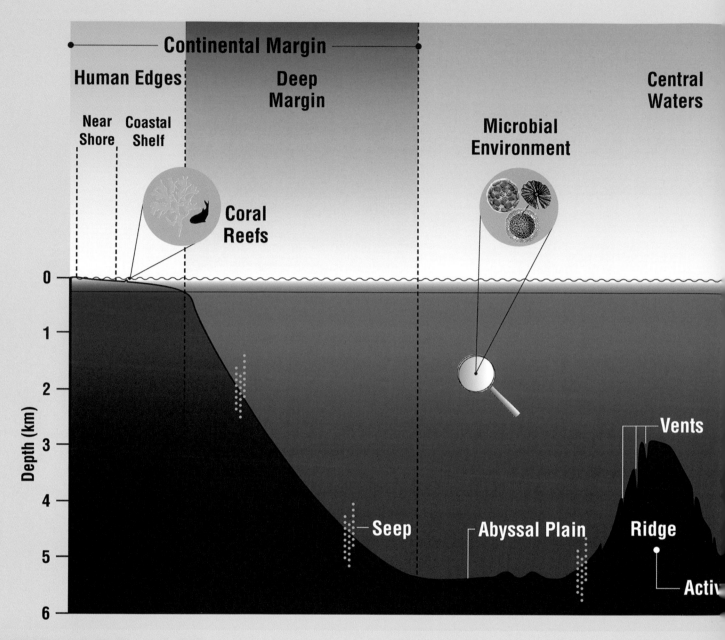

Continental Margin

Human Edges

Deep Margin

Central Waters

Near Shore   Coastal Shelf

Microbial Environment

Coral Reefs

Depth (km)

0

1

2

3

4

5

6

Vents

Seep

Abyssal Plain

Ridge

Activ

OCEAN REALMS

Because of the vast volume of ocean space, Census researchers were confronted by the challenge of separating it into sections that could be sampled and surveyed to produce meaningful results. To accomplish this nearly insurmountable task, they decided to forgo the traditional divisions of oceanic regions, and instead devised a scheme of six oceanic realms that took advantage of available exploration and sampling technology. These realms were intended to encompass all the major ocean systems and taxonomic groups in order to allow

Census scientists to document diversity, distribution and abundance. It was clear from the outset that, given the lack of records and data from depths greater than 100 meters (330 feet), a quantitative census of all realms was simply not possible, even in an ambitious 10-year program. However, it was important to proceed toward developing a baseline or benchmark from which future changes might be assessed and measured.

The Census divided the global ocean into the following realms:

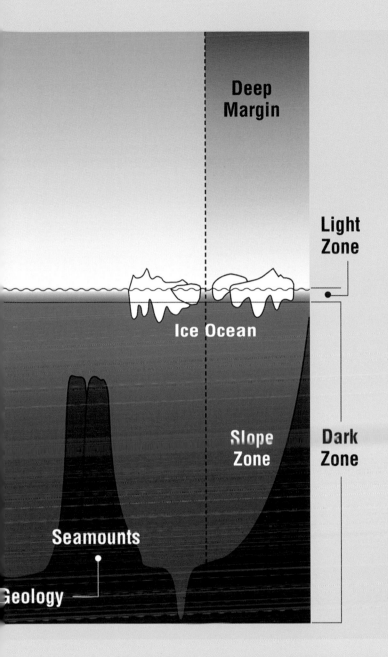

Deep Margin

Light Zone

Ice Ocean

Slope Zone

Dark Zone

Seamounts

Geology

latitudes and climates. The coastal shelf zone comprises the continental shelf and includes coral reefs.

Continental Margins and Abyssal Plains: The Census is investigating the slope region of continental margins, which begins at the edge of the coastal shelf and extends to the deep ocean basin. This area covers both the continental margins and the abyssal plains. Little prior biological data exists for these regions, which the Census refers to as "hidden boundaries" because their inaccessibility has left them largely unexplored.

Central Waters: Between the continental margins are the ocean basins and open deep water, which the Census characterizes as Central Waters. The central waters are further divided into the light zone, from the ocean surface to a depth of 200 meters (660 feet), and the dark zone, from 200 meters to the ocean floor. The light zone is home to drifters, swimmers and global travelers. The drifters — photosynthetic plants — live in the light zone and serve as a food source for the swimmers — marine zooplankton. The light zone is also home to global travelers, which are large swimming animals that traverse the ocean basin. The dark zone is similarly divided into midwater and bottom-water inhabitants. Together these large open ocean areas are home to at least 40 percent of the world's primary production of biomass.

Active Geology: Seamounts, hydrothermal vents and cold seeps are geologically active areas that the Census grouped together in the Active Geology realm. This realm has provided Census scientists with a wealth of new species.

Ice Oceans: Two Census field projects are studying ocean areas at the opposite poles of Earth: the Arctic and Southern oceans. These projects share a need for specially equipped ice-breaking ships, forcing highly integrated sampling schedules for the entire water column to get a better idea about what lives below the ice and in the frigid waters of these regions.

Microbial Environment: Microscopic marine life exists throughout the global ocean and has a dominant role in ocean processes. To complete a census of marine life, marine microbes are being studied to gain a better understanding of how microbial populations — the oldest life-form on the planet — evolve, interact and redistribute on a global scale.

Nearshore and Coastal Shelf Zones: A continental shelf is a very gradually sloping border between the edge of a continent and the ocean basin. While continental shelves comprise only 10 percent of ocean areas, they contain most of the known marine biodiversity; they also lie mainly within exclusive national economic zones. The Census defined these "human edges" to be from the high-tide line to the bottom of the continental shelf, and divided them into nearshore and coastal shelf zones. The nearshore was defined as between the high-tide line and a water depth of 10 meters (33 feet); it stretches for more than a million kilometers around the global ocean and across

The Nearshore

Census researchers use standardized research methods in nearshore areas across all latitudes, climates and ecosystems so that results can be compared and contrasted. This linkage is being accomplished through the field project Natural Geography in Shore Areas, or NaGISA (*nagisa* means "coastal environment" in Japanese); its goal is to assess, visualize and explain nearshore biodiversity patterns. More than 120 sampling sites encompassing three-quarters of the world's coastlines have been established.

The Coastal Zone

Recognizing that many marine species migrate over long distances, another coastal project, Pacific Ocean Shelf Tracking (POST), is building a permanent acoustic array along the west coast of North America to track juvenile Pacific salmon and other species as small as 10 grams. Fish are implanted with acoustic tags that emit individually unique signals, which are picked up by listening arrays placed along the coast. This initial tracking system has served as a prototype for a larger, worldwide array that will track a spectrum of animals from squid to eels to whales, to be implanted with acoustic tags by 2010.

Recent crises in fisheries have forced a reexamination of how single species are managed. New management strategies are evolving, in part because of one Census field project, the Gulf of Maine Area (GoMA) program, which is identifying and collecting biological knowledge necessary for ecosystem-based management in a large marine environment. It is hoped that this project will serve as a model for future efforts of this kind by providing adequate data and understanding for better management of fishery resources around the globe.

Coral Reefs

Scientists estimate that less than 10 percent of what lives on coral reefs has been identified. The Census of Coral Reefs (CReefs) field project was launched to help fill this knowledge gap. Researchers linked across latitudes and climates are using standardized research methods to explain diversity patterns before those patterns are further affected by global changes.

These sea stars were found near the shore of Cobscook Bay, Maine, in August 2000.

Continental Margins

The Census is investigating the slope region of continental margins, which begins at the edge of the coastal shelf and gradually angles down to the deep ocean basins. Because of their sloping topography, distance from shore and depth, these areas have been little studied. To explore these "hidden boundaries," a Census project called Continental Margin Ecosystems (COMARGE) is working with oil companies to establish biodiversity baselines in margin areas worldwide.

Abyssal Plains

The deep-sea floor below the base of the continental slope covers approximately 40 percent of Earth's surface, which is more than the total surface mass of the continents (29 percent). This region, from approximately 4,000 to 6,000 meters (2.5 to 3.75 miles) in water depth, has large,

Photographed at Kingman Reef in the Line Islands, this anemone, Heteractis crispa, *is found throughout the tropical Pacific, everywhere except Hawaii. It lives in both calm lagoons and channels with swift moving current, and varies in color from purple to tan. An opportunistic feeder, it traps fish, crabs and other reef organisms in its tentacles and passes them to its centrally located mouth. But it also serves as a home to clown fish or damselfish that often live unharmed among its tentacles.*

This blind lobster with bizarre chelipeds (claws) belongs to the very rare genus Thaumastochelopsis, *previously known only from four specimens of two species in Australia. This specimen, collected from a depth of about 300 meters (1,000 feet) in the Coral Sea, is a new species.*

relatively flat areas called abyssal plains, where very little is known about the inhabitants. The Census of Diversity of Abyssal Marine Life (CeDAMar) field project was launched to investigate the biodiversity of species living in, on or directly above the sediment in these regions. The first two expeditions found hundreds of new species, and five more expeditions are planned by 2010. It is likely that CeDAMar will make one of the largest contributions to the database of known marine species, without exhausting the potential for additional new discoveries.

THE LIGHT ZONE

Drifters and Swimmers The foundation of life on land is the photosynthetic activity of plant life. Similarly, the foundation of ocean life is the photosynthetic activity of "plants" in the sea: phytoplankton, photosynthetic bacteria and algae. Phytoplankton, comprising almost 95 percent of total marine productivity, live in the top 200 meters (660 feet) of the global ocean, where they find the sunlight needed to photosynthesize.

New technology has given Census researchers an insider's view of the small animals that inhabit the parts of the world's ocean where light penetrates. Shown here is an amphipod, Eusirus holmii, *as it searches for small prey to seize with its powerful claws.*

Phytoplankton are consumed by diverse populations of zooplankton, which are dominated by species of crustaceans called copepods. The Census of Marine Zooplankton (CMarZ) is attempting a global-scale analysis of all marine zooplankton species, using new and emerging technologies that include molecular, optical and acoustical imaging, and remote detection.

Global Travelers The light zone is also home to large swimming animals that travel across ocean basins. Scientists are learning about these animals through tagging and real-time tracking. A Census field project, Tagging of Pacific Predators (TOPP), is working with marine animals to create a view of the vast open-ocean habitats as seen by the animals themselves. The travels of top predators are especially intriguing. TOPP scientists have tagged 23 species — more than 2,000 animals — from albatross to albacore tuna to elephant seals and squid.

The Dark Zone: Mid-waters and Bottom Waters

The challenge of investigating the dark zone – water below 200 meters (660 feet) – was undertaken by the multinational group of researchers who formed the Mid-Atlantic Ridge Ecosystems Project (MAR-ECO). Their goal was to explore and understand the distribution, abundance and trophic (food) relationships of the organisms inhabiting the middle and deep waters of the mid-oceanic North Atlantic.

A multidisciplinary trans-Atlantic team of researchers has used ships and submersibles to carry out investigations between Iceland and the Azores. One two-month investigation in mid-2004 along the mid-Atlantic Ridge was the most comprehensive survey of this realm to date, both qualitatively and quantitatively. Sampling yielded 45 to 50 squid species (two potentially new to science) and 80,000 fish specimens, many of which are thought to be new to science, or at least new to the North Atlantic. The team is leading efforts to extend this new technology to other ocean basins.

Active Geology

Hydrothermal Vents and Cold Seeps Hydrothermal vents are places on the ocean floor where a continuous flow of superheated, mineral-rich water flows up through cracks in the Earth's crust. Cold seeps are areas of the ocean floor where hydrogen sulfide, methane and other hydrocarbon-rich fluids, the same temperature as the surrounding seawater, slowly seep from the seafloor. These geologically active areas are being investigated by the Census field project Biogeography of Chemosynthetic Ecosystems (ChEss). ChEss is discovering new vents and seeps, primarily along the Equatorial Atlantic Belt, in the southeast Pacific and off the coast of New Zealand, and the scientists are establishing a global database of the species they find. New species have been discovered at the rate of about two per month since the first hydrothermal vent was found in 1977.

Seamounts Seamounts are typically steep-sided extinct volcanoes that lie below the ocean surface. An "official" seamount must be at least 1,000 meters (3,300 feet) high. The peaks of these undersea mountains are usually found from a few hundred to a few thousand meters below the sea surface. The Census of Seamounts (CenSeam) has conducted several expeditions to many previously rarely explored areas to learn more about what lives on and near different types of seamounts and why species may

Opposite: An amazing array of exquisitely beautiful animals, such as this jeweled squid (Histioteuthis bonelli)*, has been found in the previously little-explored dark zone of the world's ocean.*

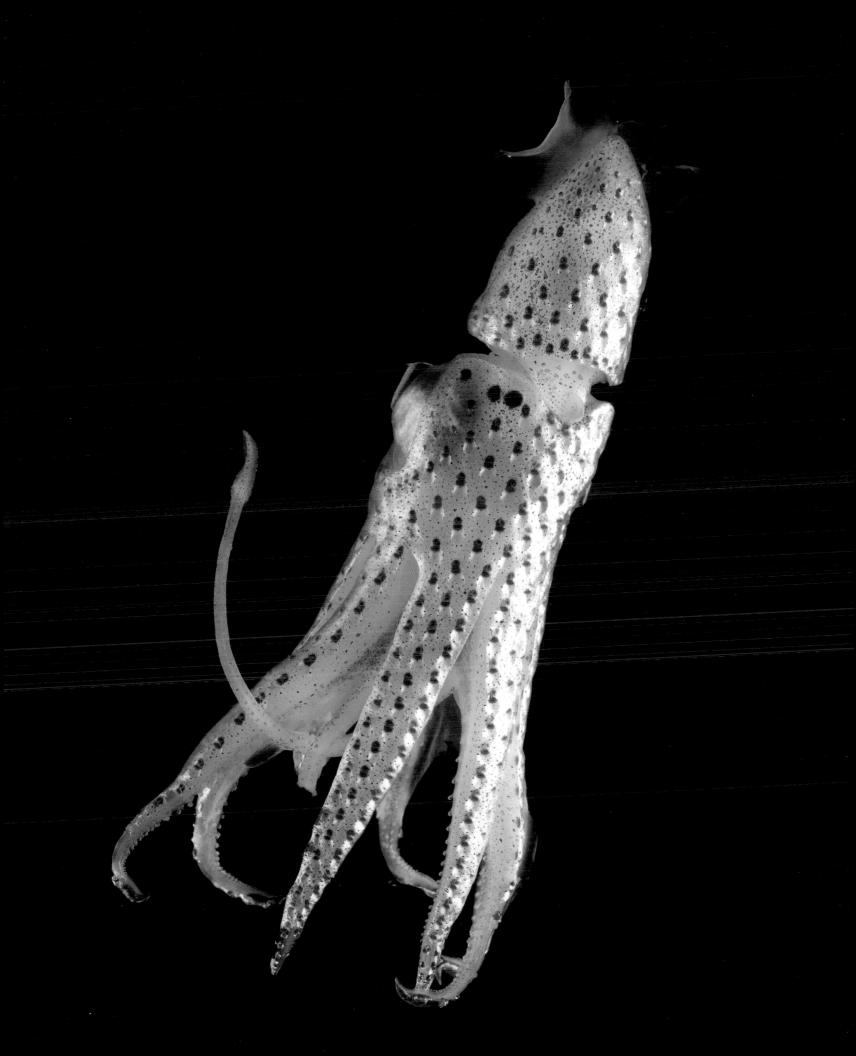

A surprisingly rich diversity of species is being discovered in areas such as hydrothermal vents, the first of which was discovered in 1977. This "black smoker" vent is at Logatech, on the Mid-Atlantic Ridge.

Census scientists are finding that richly diverse communities inhabit seamount areas around the world. This is a community of orange roughy fish (Hoplostethus atlanticus).

thrive in these unique habitats. Scientists are also investigating the impact of commercial fisheries on seamount communities.

ICE OCEANS

Two Census field projects are studying the ocean at Earth's poles. To do so, they require specially equipped ice-breaking ships and highly sophisticated sampling gear to study the entire water column from surface ice to the ocean bottom. The Arctic Ocean Diversity (ArcOD) project's goal is to assemble existing knowledge of biodiversity in this least-known ocean, direct new international explorations using new technologies, and create a framework for understanding and predicting biological changes in an ocean that contains rapidly melting ice.

Sampling in the Arctic Ocean is most often done during the summer months but still requires the services of ice-breaking ships to reach little-explored areas.

Loss of ice in the Antarctic Peninsula has opened up new areas for exploration by Census scientists to study how life adapts in cold, remote Southern Ocean regions.

Like its northern counterpart, the Census of Antarctic Marine Life (CAML) project is collecting rich biological data on the Southern Ocean and encouraging biodiversity sampling by all cruises that visit the region. CAML led 18 expeditions during the International Polar Year that ended in the spring of 2008. The goal is to couple this new understanding of biology with the complex current dynamics that control life throughout the global ocean.

Microbes

The number of microbes in marine environments is staggering. Census researchers have found more than 20,000 microscopic organisms in one liter (1.1 quart) of seawater. The International Census of Marine Microbes (ICOMM) project is developing a biodiversity database for marine microbes to help understand how microbial populations – the oldest life-forms on the planet – evolve, interact and redistribute on a global scale. What is learned may potentially redefine our understanding of biodiversity from the microbial level to that of the large marine mammals that traverse the global ocean.

The Antarctic ice fish is an example of how life adapts below a collapsed ice shelf in the Southern Ocean. This fish has no red blood cell pigments (hemoglobin) and no red blood cells. Because its blood is thin, it can save energy otherwise needed to pump thicker blood through its body.

A GALLERY OF MICROBES

Microbes are intricately complicated, varied and in many instances quite beautiful. These microbes were photographed with an electron microscope.

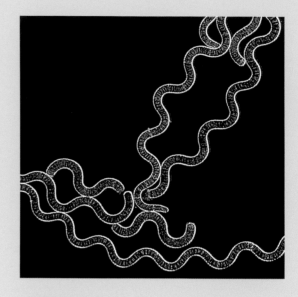

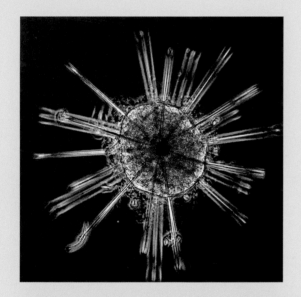

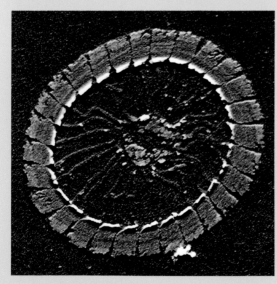

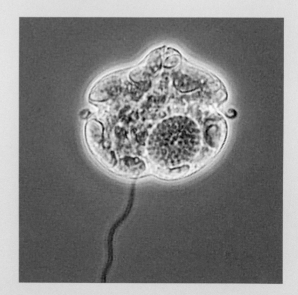

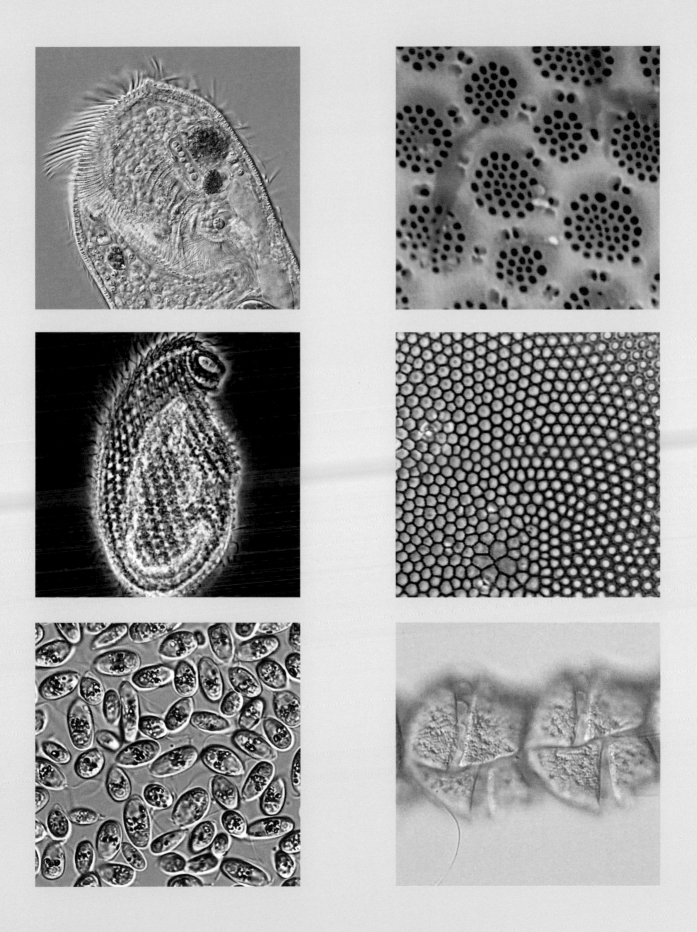

CHAPTER TWO

Painting a Picture of the Past

To understand the present and begin to predict the future,
one must first understand the past.

— Patricia Miloslavich, co-senior scientist, Census of Marine Life,
Universidad Simón Bolívar, Caracas, Venezuela

In any endeavor, placing the present situation in context and trying to predict what the future may hold are impossible without an understanding of the past. Historical context, like memory, frames current conditions and can suggest future scenarios and trends. Census scientists understood this from the outset, but they faced many challenges in designing a research program that could fairly, accurately and credibly determine what once lived in the world's ocean. To answer this question, they launched a study of the history of marine animal populations.

WHAT ONCE LIVED IN THE OCEAN

What once lived in the ocean? Finding an answer demanded a new perspective that took a broad view of the marine sciences and used a multidisciplinary approach. After considerable negotiation, fisheries scientists, marine historians, marine ecologists and other experts in related fields established a time frame for assessing the history of marine animal populations, focusing primarily on the past 500 years. This period corresponds

Opposite: Winslow Homer,
American, 1836–1910
The Fog Warning, *1885*
Oil on canvas
76.83 x 123.19 cm
(30¼ x 48½ inches)
Museum of Fine Arts, Boston
Otis Norcross Fund, 94.72

to the beginning of European expansion into Africa, Asia and the Western Hemisphere, which was soon followed by the earliest global expansion of commercial fishing. It is widely accepted that humans affected the marine environment well before 1500, but record keeping improved considerably after the 16th century, so the modern-day investigators turned to archival records to find historical data on fish and fisheries. Instead of restricting their research to the original 500-year time frame, however, the marine scientists, archeologists, paleozoologists and other researchers analyzed a wide variety of evidence related to fishing and other human use of ocean resources, some dating back thousands of years, which added to their understanding of how fishery resources changed over time.

Shell middens in Blombos Cave, South Africa, possibly dating back some 140,000 years, document that human harvesting of ocean animals came before the development of agriculture and cities. Murals from ancient Egypt, Crete and the Mayan Yucatán and the whale-ivory fish-hooks and oral traditions of Inuit and Polynesian peoples are only a few of the countless examples that illustrate how the sea contributed to the prosperity of early civilizations. Nevertheless, neither humans' impact on marine ecosystems nor the role of ocean resources in early human development has been widely studied. Until recently, marine ecologists generally used data collected after the Second World War to assess the current state of commercially fished marine species. Historical data were widely ignored because they did not conform to rigorous scientific standards.

WHAT IS NATURAL?

In 1995, fisheries biologist Daniel Pauly of the University of British Columbia coined the term "shifting baselines" to describe how standards defining what is normal can degrade over time. Pauly suggested that, because people gauge what is "natural" according to their own experience, successive generations of marine scientists had overlooked changes that had taken place in the oceans before their time. Knowledge about the natural environment of the past existed only in stories and anecdotes, not as the quantifiable data arrayed over time that scientists usually rely upon. Thus, with each subsequent generation of investigators, the baseline conditions for "normal" shifted ever further into unnatural territory.

Baselines of ecosystem health are important reference points because they represent standards against which change can be measured. These standards are critical in determining the magnitude, rate and direction of change in ocean ecosystems over time and the degree to which change is a result of human activities, as well as the extent of exploitation and the impact of climate change on global marine resources.

In 2001, Jeremy Jackson of the Scripps Institution of Oceanography brought the reality of shifting baselines to the public's attention, when he and 18 co-authors published the results of their analysis of historical data on marine life in coastal oceans around the globe. Jackson and his team concluded that overfishing had been widespread for over a millennium: in the Caribbean, in Chesapeake Bay, off California and Australia – everywhere. It is virtually impossible to imagine how much the oceans of the past teemed with life.

These landmark papers changed the way people thought about the ocean, and created a new desire to understand its history. A historical perspective became an integral part of the Census work. By 2003, researchers were engaged in the arduous compilation of long historical data sets about marine animal populations, their goal being to distinguish natural fluctuations from the effects of human activities. Ultimately, the results of the 16 historical case studies undertaken by the Census will present the first reliable picture of life in the ocean both before and after human impact became significant.

The historical component of the Census of Marine Life set out to improve understanding of ecosystem dynamics by investigating long-term changes in the abundance of marine life; the ecological impact of continuous, large-scale human harvesting; and the role that marine resources played in the development of human society. A secondary objective was to show that quantitative in-depth analysis of the healthier marine species and ocean environments described in the historical documents could improve goals for marine management in the present and promote recovery in the future. This agenda required a multidisciplinary approach that integrated marine ecology, maritime history, archeology and paleoecology to create a synthesis of research methods and analytical perspectives. No study of this scope and magnitude had ever been undertaken, and the scientists quickly discovered that a project this large presented many obstacles, not the least of which was finding data suitable for mathematical modeling.

Whaling Logs, Menus and Other Records

Since the written records of historical fish populations were mostly limited to species with commercial value, researchers initially focused their efforts there. The first tasks were to identify, recover, transcribe and interpret data sets pertaining to marine species from historical records. Some sources were foreign to marine science and had never been used or linked together in this way before. The documents ranged from Russian monastery records of the 1600s and dusty old whaling logs stored in archives around the world to Australian landing records from the 20th century, early scientific survey cruises starting in the 19th century, and tax records from the Baltic region. In one instance, American restaurant menus served as a source, not only for changing dietary preferences but also as an indicator of the availability and stock abundance of marine species.

Non-written sources were as varied as fish bones recovered from archeological digs of medieval England and Scotland to paleoceanographic data in Estonia, where archeologists found imported cod bones dating from the late 13th century. All these diverse sources

Primary sources used by Census investigators included the meticulously kept log books of whaling expeditions.

Census researchers used a wide variety of resources to track the past. Shown here are some of the 200,000 restaurant menus that served as primary sources for tracking the vagaries of seafood and fish offerings – and pricing – over time.

Left: As fishing methods have changed, improved techniques, in many instances, have had harmful effects on fish populations. Shown here is a weir for catching salmon on the Kitsa River, Russia, circa 1850.

Below left: Halibut have virtually disappeared from the North Atlantic because of overfishing, and any that are caught now are much, much smaller than the giant shown here in a circa 1910 postcard showing a large halibut (123 kg/270 lb.) caught off Provincetown, Massachusetts.

Below: Census researchers are finding that the size and availability of many marine species are different from those of the past, and in many cases individual catches are smaller than in the past. This large cod was caught off Mohegan Island, Maine.

A 270 LB. HALIBUT CAUGHT AT PROVINCETOWN, MASS.

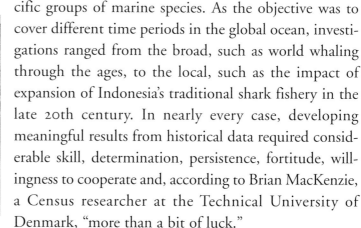

This whale was killed by Capt. Joshua Nickerson in the steamer A. B. Nickerson, and was one of the largest of the Finback species ever taken here and measured as follows:

Length, 65 feet, 4 inches
Across the tail, 14 feet, 6 inches
Length of lower jaw, 11 feet
Length of fins, 10 feet
Girth, 37 feet
Weight, 136 tons
Capacity of mouth when closed, 30 barrels.

WHALE ASHORE ON BEACH AT PROVINCETOWN, MASS.

Old postcards help document the past history of fisheries in various regions. These show whaling of both large and small species off Cape Cod around 1900. It has been largely forgotten that whaling took place in Cape Cod Bay and near Stellwagen Bank just 100 years ago.

were analyzed to create a time series of stock abundance and geographic distribution that was weighted with information about the influence of fishing, climate variability and other factors that may have caused changes in marine animal populations.

Researchers investigating the history of marine animal populations were organized into teams that focused on a particular region or on spe-

Black Fish driven ashore at South Wellfleet, Cape Cod, Mass. About 1500 in the school. Sold for fifteen thousand dollars, which was divided among 300 inhabitants.

cific groups of marine species. As the objective was to cover different time periods in the global ocean, investigations ranged from the broad, such as world whaling through the ages, to the local, such as the impact of expansion of Indonesia's traditional shark fishery in the late 20th century. In nearly every case, developing meaningful results from historical data required considerable skill, determination, persistence, fortitude, willingness to cooperate and, according to Brian MacKenzie, a Census researcher at the Technical University of Denmark, "more than a bit of luck."

Census researchers responded to that challenge and are reporting the results of up to eight years of work as individual case studies are completed. A major synthesis of all the results will be prepared for the first Census in 2010. The research methods and results presented here demonstrate the complexity of the work and the potential value of its findings.

RECORDS OF DECLINE

When scientists uncover the ocean's past and compare it to its present, the results are often shocking. For example, new Census findings suggest that northern Europe's Wadden Sea has been experiencing environmental decline for up to a thousand years longer than previously thought.

By analyzing and synthesizing archeological and historical data on social and ecological changes, Census researchers led by Heike K. Lotze of Dalhousie University in Halifax, Nova Scotia, constructed a timeline of human-induced environmental decline in this shallow portion of the North Sea. Bordering Denmark, Germany and the Netherlands, the Wadden Sea has historically been an important source of food, transportation and other resources for the local population. While it is not surprising that human use has caused environmental and ecological decline, these new findings indicate that degradation has been taking place for much longer than scientists formerly thought.

In an attempt to set a baseline for the Wadden Sea's "natural" state, researchers examined the historical record to see how long humans have been exploiting this body of water. Such a baseline would reveal the extent of degradation in the ecosystem and could prove a valuable tool for management of this heavily used coastal area. During the analysis, researchers were surprised to find records of overexploitation of marine resources going back at least 500 years, rather than the commonly accepted figure of 150 years. Although eutrophication – an influx of excess nutrients such as nitrogen and phosphorus – began about 100 years ago, records show that human-driven environmental change extends back at least 1,000 years to when coastal populations began building dikes to reclaim land from the sea.

As a result of this study, the baseline time period for a "natural" Wadden Sea has shifted backward from its previous projection, around 1850, to about 1000 AD, before the Norman conquest of England and far beyond the scope of current monitoring programs and management strategies. The implication is that current regulatory programs are comparing changes in the environment to baselines that have been set for the wrong period. As monitoring and management efforts evolve in the future, these findings will reveal to scientists, policymakers and the public a more complete picture of human influence on the Wadden Sea. In turn, new targets for restoration and recovery may be developed and implemented, targets that take into account the entire transformation undergone by this ecosystem over a thousand years.

Human exploitation of marine resources has occurred throughout history. Depicted in this line drawing is a sturgeon slaughter in Hamburg Harbor, Germany, in the late 19th century.

Another historical study led by Heike Lotze examined historical changes in estuaries – where rivers flow into the sea – around the globe. Researchers combined paleontological, archeological, historical and ecological records to trace changes in important animal populations, habitats, water quality parameters and species invasions. The results, from 12 major estuaries and coastal seas around the world, painted a picture of humans' impact on coastal marine ecosystems from Roman times to the present. Lotze and her colleagues concluded that human activities over time have depleted 90 percent of important marine species, eliminated 65 percent of sea-grass and wetland habitat, degraded water quality ten- to a thousandfold, and accelerated species invasions. Moreover, the decline has accelerated in the past 150 to 300 years as populations have grown, demands for resources have increased, luxury markets have developed and industrialization has expanded.

"Throughout history, estuaries and coastal seas have played a critical role in human development as a source of ocean life, habitat for most of our commercial fish catch, a resource for our economy and a buffer against natural disasters," explains Lotze. "Yet these once rich and diverse areas are a forgotten resource. Compared to other ocean ecosystems such as coral reefs, they have received little attention in the press and are not on the national policy agenda. Sadly, we have simply accepted their slow degradation."

Most of the mammals, birds and reptiles in these 12 estuaries were depleted by 1900, and by 1950 they had declined further with the growing demand for food, oil and luxury items such as furs, feathers and ivory. Fish species such as the highly desirable and easily accessible salmon and

Opposite: This satellite photograph shows the devastating effects of Hurricane Floyd (1999) on the Outer Banks of North Carolina. Green indicates vegetation on land and phytoplankton in the water.

Humans have played a huge role in the vitality of marine ecosystems. Greek sponge divers, for example, totally changed the Florida sponge fishery with their advanced diving technology, which allowed them access to pristine previously unfished areas. These divers were photographed in Tarpon Springs, Florida, circa 1931.

sturgeon were depleted first, followed by tuna and sharks, cod and halibut, and herring and sardines. Oysters, because of their value and accessibility as well as destructive harvesting methods, were the first invertebrate resource to decline. The primary cause of estuarine damage is human exploitation, which is responsible for 95 percent of species depletions and 96 percent of extinctions, often in combination with habitat destruction. In the future, however, invasive species and climate change may play a larger role in reducing these already depleted resources.

Fortunately, Census research has uncovered some good news. In estuarine areas where conservation efforts were implemented in the 20th century, signs of recovery are appearing. According to Census scientists, the fastest path to recovery has been through moderating the cumulative impacts of human activity. Seventy-eight percent of recoveries have been achieved by reducing at least two human activities at once, for example, resource exploitation, habitat destruction and pollution. Trends suggest that estuaries in developed countries may have passed their low point and are on the path to recovery. However, population growth in developing countries continues to puts pressure on estuaries in their coastal areas that may increase their degradation and hamper further mitigation efforts. But Census research has given scientists a valuable tool: a historical baseline that allows them to understand what coastal ecosystems looked like before human interference, and that will also serve as a vision for regenerating resilient ecosystems that can thrive in the years ahead.

IN QUEST OF A ZERO-YEAR BASELINE

Another Census investigation is using direct sources of evidence – ancient fish bones from archeological digs – to identify the earliest human impacts on abundant marine species, with the goal of setting a "year zero" for shifting baseline studies. Protein from fish bones recovered from archeological digs is providing insight into the fishing practices of civilizations of long ago. As James Barrett, head of the fish-bones project at the University of Cambridge, explains, "One of our main objectives is to chart the ebb and flow of fishing, using bones from archeological sites." The scientists' ultimate goal is to figure out if they can accurately define how long humans have been influencing marine ecosystems in northern and western Europe. The work raises many questions, including how far back one should go and what scale should be studied in order to find the origin of a given fishery. Basic questions such as what species were being fished, who was fishing, when and where they were fishing and for what market also pose many challenges to creating an accurate history of fishing practices long since past.

To add to the information provided through analysis of what fish were caught and traded, scientists turned to the exacting process of measuring stable isotopes in the fish-bone protein (isotopes are naturally occurring forms of the same element that have different numbers of neutrons in their atomic nuclei). Protein survives in bones even after thousands of years, so ancient fish bones can tell us what the fish ate and what ate them; scientists call this their trophic history. Since the bigger fish at the top of the food web carry a different isotopic signature than the smaller ones below, the characteristics of protein present in a fish bone allows scientists to determine where the fish fell in the food web.

James Barrett and his team of researchers are using these techniques to explore when various societies "fished down" the food web, where the fish were caught and whether early eutrophication might have occurred. By learning how people fished in the past, scientists hope to arrive at a better understanding of both natural ecosystem fluctuations and human-induced changes in specific regions.

THE IMPACT OF GLOBAL WARMING

To understand how global warming may be affecting marine fish species, Inge Bødker Enghoff, of the Natural History Museum of Denmark and

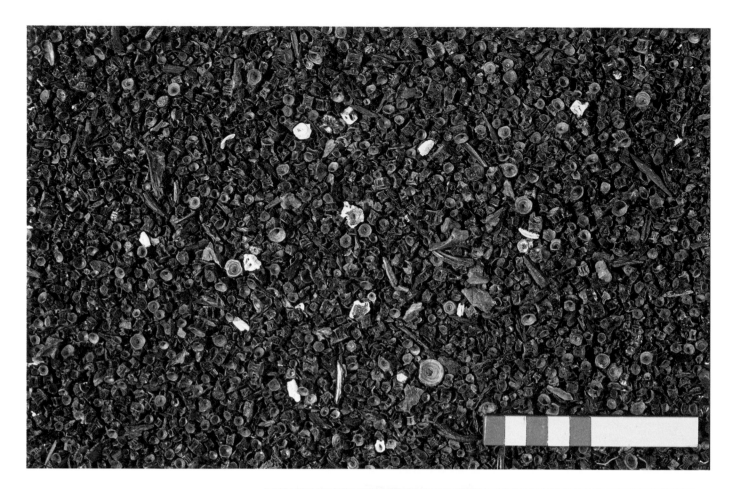

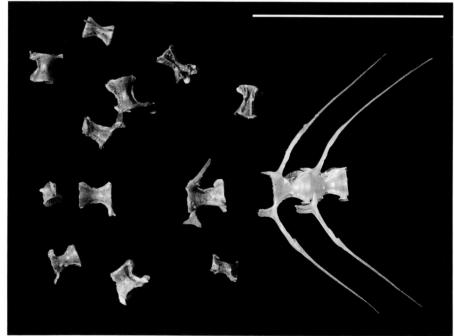

Above: This photograph shows just part of the fish bones obtained from 1 square meter of the Maglemosegaard excavation in Denmark. Approximately 48 percent of the 12,784 bones are from gadids, mainly cod. (The red bars on the scale are 1 centimeter/0.4 inch wide.)

Right: These vertebrae are from an anchovy caught by Stone Age people at Krabbesholm, Denmark; two connected vertebrae are shown for comparison. (The scale line is 1 centimeter/0.4 inch long.)

the University of Copenhagen, teamed with Census colleagues Brian MacKenzie and Einar Eg Nielsen from the Technical University of Denmark to investigate fish fauna during one of the warmest prehistoric periods: the warm Atlantic period, circa 7000 to 3900 BC. Working with 108,000 fish bones dating from that period, taken from Danish archeological deposits, the researchers found several species – for example, anchovy and black sea bream – that were typical of waters much farther south and much warmer, such as the Bay of Biscay and the Mediterranean Sea. These species disappeared from the archeological record after temperatures cooled, but researchers have seen them reappear in commercial landings and research vessel survey data during the past 10 to 15 years as temperatures have risen in the waters around Denmark. The re-emergence of warm-water species may prove that archeological information can help predict future species' composition as climate change progresses and temperatures rise.

MANAGING THE FISHERIES

Not only are innovative multidisciplinary research techniques providing insight into historic marine animal populations, they may also have lasting implications for present and future fishery resource management. One example is the work of Census investigator Andrew A. Rosenberg and his colleagues at the University of New Hampshire. They embraced the daunting challenge of documenting the decline of cod populations off the coast of Nova Scotia, and found that past stock levels may have significance for current management initiatives. Using catch records and observations in 19th-century fishing logs, coupled with modern modeling tools, Rosenberg's team compiled the earliest estimate of cod population abundance on the Scotian Shelf. Although they were once a dominant species in that ecosystem, the biomass of cod has plunged 96 percent. Just 16 small pre-Civil War schooners could hold all the adult cod estimated to be living on the Scotian Shelf in 2005.

Using a modified mathematical model developed for fish-stock assessment, the researchers estimated that in 1852 the cod biomass on the Scotian Shelf was 1.26 million metric tons, compared with less than 50,000 metric tons today. Rosenberg stresses that this estimate is actually "quite conservative." Furthermore, since the size of the hooks used in the 1850s made it very unlikely that smaller juvenile cod were landed, most of the cod landed by the fishing schooners were adults. Adult cod today are estimated to weigh 3,000 metric tons and make up only 6 percent of the total cod biomass on the Scotian Shelf.

Rosenberg suggests that this 150-year perspective challenges conventional wisdom as to what constitutes a rebuilt cod stock in a productive marine environment. The best estimate of adult cod biomass on the Scotian Shelf today comprises only 38 percent of the catch brought home by 43 schooners from Beverly, Massachusetts, in 1855. As attempts are made to rebuild the cod and other fisheries, Rosenberg suggests that the past potential of the stock must be considered in setting present-day management goals.

CHARTING A NEW COURSE OF STUDY

New ways of conceptualizing and investigating historical fisheries have led to the creation of a new field of study. As a result of Census scientists' groundbreaking work, three new centers for marine environmental history have been established: at Roskilde University in Denmark, at the University of New Hampshire in the United States and at the University of Hull in the United Kingdom. These institutions serve as central

This pile of abalone shells on a circa 1920 postcard illustrates the reality of shifting baselines.

ABALONE SHELLS, SANTA BARBARA, CAL.

coordinators of the historical arm of the Census of Marine Life; they maintain research focus; identify, aid and implement priority research projects; and ensure that the individual studies are synchronized. The Centers also train graduate students in the multidisciplinary methods of ecological, historical and paleoecological research so that the richness of the past may influence management decisions in the present and help guide future policy on the oceans.

Undoubtedly the Census of Marine Life has expanded knowledge of past life in the global ocean and of the factors that control marine animal populations. Its work has increased understanding of the role of marine resources in human history. By displaying the advantages of a multi-disciplinary team for tackling complex scientific questions, its comprehensive approach has broadened the way science is practiced. Perhaps most important, historic baselines for marine species and ecosystems developed by the Census may assist in managing the recovery of threatened resources by setting responsible goals based on healthy levels of abundance, distribution and biocomplexity in the past. Finally, history reveals the many different ways in which people lived near, worked on and benefited from the ocean. In this era of unprecedented environmental change, such cultural models offering hopeful alternatives to harmful practices become increasingly important.

UNLOCKING THE MYSTERY OF THE DISAPPEARING TUNA

Brian MacKenzie, a Census researcher at the Technical University of Denmark, focused his research on historic tuna populations in various parts of the North Atlantic, which ultimately turned him into a private detective. Rather than working in a traditional laboratory or field setting, MacKenzie spent an enormous amount of time in libraries, government buildings and other document repositories. Wading through catch records, newspaper clippings and old sportfishing magazines, he discovered that abundant bluefin tuna populations once existed in places where they are no longer found today.

MacKenzie's investigation into the history of bluefin tuna was initially prompted by happenstance and then fostered by an insatiable curiosity. "I was sitting in a library one day and came upon a book about tuna in Danish waters written in 1949, when commercial and sportfishing of tuna took place. I started to read it out of curiosity more than anything else and discovered that there was a story there that really needed to be told. The book reported on bluefin tuna that were up to 200 kilograms in weight and 2 meters long – none of which are seen off the Danish coast today."

His first foray into the library turned into an extensive investigation and review of dusty old government reports and early scientific papers about fisheries based on catch data. After about 1937, the Government of Denmark required reports of commercial landings and kept statistics based on minimum landing information, which occasionally included some size data. After exhausting the supply of Danish records, MacKenzie

Tuna were plentiful in northern European waters during the first part of the 20th century, and European market halls were filled with sizable tuna for auction. In the foreground are 11 bluefin tuna caught by German fishermen in a single day in 1910.

Northern European waters teemed with majestic Atlantic bluefin tuna up until the late 1950s, as evidenced by this successful outing.

expanded his search to other countries, including Norway, where fisheries landings were five to ten times bigger than in Denmark. The Norwegian government had more extensive data that included length, weight and age information – based on the number of vertebrae rings – up until the mid-1960s, when the bluefin tuna fishery went into a noticeable decline.

After focusing on information in formal and informal government reports from the late 1930s to the early 1950s, MacKenzie moved on to older, less formal filings, such as reports from commercial fisheries' associations. He pulled together a patchwork of facts about where, when and how tuna congregated and were subsequently caught. Old taxonomic guides were sources for the years and months when bluefin tuna were seen in the waters of northern Europe – the North Sea, the Norwegian Sea and the straits Skaggerak, Kattegat and Øresund. Dates and references such as "tuna seen washing up on beach," "fishermen seeing schools in specific locations" and "tuna being caught as

bycatch in herring traps near shore" were matched with more precise scientific and government reports to draw a historical picture of the rise and fall of bluefin tuna in northern European waters.

MacKenzie teamed up with a Census colleague, the late Ransom A. Myers of Dalhousie University, who was investigating the future of marine animal populations. Combining their respective areas of research and using statistical modeling, MacKenzie and Myers published a paper in 2007 describing the abundance of the bluefin tuna population in northern European waters in the early 1900s. The fish typically arrived in northern waters by the thousands in late June and departed by October at the latest, their foraging travels likely related to seasonal warming.

In the 1920s an industrialized fishery began to gear up. The number of bluefins caught increased considerably, which fueled demand for these tasty fish. Better means of catching tuna also encouraged sportfishing and its associated businesses. Catches were plentiful and the sporting pleasurable – in

1928 one sport fisher was reported to have caught 62 bluefins in one day near the Danish island of Anholt. Such easy landings generated enthusiasm among sport fishers, and the news spread quickly through the United Kingdom, Norway and elsewhere in northern Europe. An example of the new popularity of this once humble fishery was the establishment of the Scandinavian Tuna Club, which arranged recreational bluefin tournaments in the straits between Denmark and Sweden up until the early 1960s.

Booming catches, both recreational and non-recreational, helped deplete the Atlantic bluefin population in a single generation. An explosion of heavy fishing from the late 1930s to the mid-1960s ultimately led to the fishery's collapse and the disappearance of tuna from these waters. Explains MacKenzie, "We can't say with certainty that over-exploitation is the smoking gun in the bluefin tuna's disappearance – but clearly there was a murder."

MacKenzie and his colleagues were prompted to widen their investigation. They found that the rise and fall of bluefin tuna fisheries in northern European waters were replicated elsewhere: fisheries developed and crashed off the coast of Brazil and northern Argentina in the late 1950s and early 1960s, and the fishery in the Black Sea ended in 1986. These findings prompted MacKenzie to carry out further research. Today he deciphers how fishing and environmental variables, such as sea temperature, food abundance and water circulation, may affect the presence or absence of tuna in specific areas that may influence migratory and feeding behaviors. The goal of his team is to figure out exactly what happened to the bluefin tuna population so that changes, if needed, can be made to prevent such population declines from occurring again.

Tuna were so abundant in northern European waters after the First World War that an active sportsfishery developed.

Tuna Timeline

Before 1910

Bluefin tuna were rarely captured in northern European waters before 1910, and even coastal sightings were exciting events. One fish measuring 2.7 meters (almost 9 feet) washed up on a German shore in 1903. However, archeological evidence shows that tuna were caught in Denmark and Norway during the 1400s to 1600s and even thousands of years ago, circa 7000–3900 BC.

1910 to 1920s

Ever better know-how and equipment, including harpoon rifles and hydraulic net lifts, helped northern European fishers land burgeoning quantities of bluefin. In 1915 nearly 8,000 bluefin (690 tons) were landed in Gothenburg, Sweden, alone. Those captured in the 1920s ranged from 40 kilograms to giants of 700 kilograms (90 to 1,500 pounds), with an average weight of 50 to 100 kilograms (110 to 220 pounds). In the 1920s the catch also peaked at Boulogne, home port for French bluefin fishers in the North Sea. In 1929 Denmark built its first tuna cannery, a milestone in the new industrial approach to processing fish.

1940 to 1950s

Tuna-fishing countries such as Norway, Denmark, Sweden and Germany recorded virtually no bluefin landings in 1910, but by 1949 they were collectively reporting landings of almost 5,500 tons. Landings of bluefin tuna in record numbers by northern European boats continued throughout the 1940s, and by decade's end they approached the catch levels of traditional Mediterranean fisheries. In 1949 Norway had 43 boats in pursuit of the bluefin; the next year it had 200. Norwegian catches briefly exceeded 10,000 tons per year in the early 1950s.

Tuna were large, plentiful and frequently caught without sophisticated fishing gear.

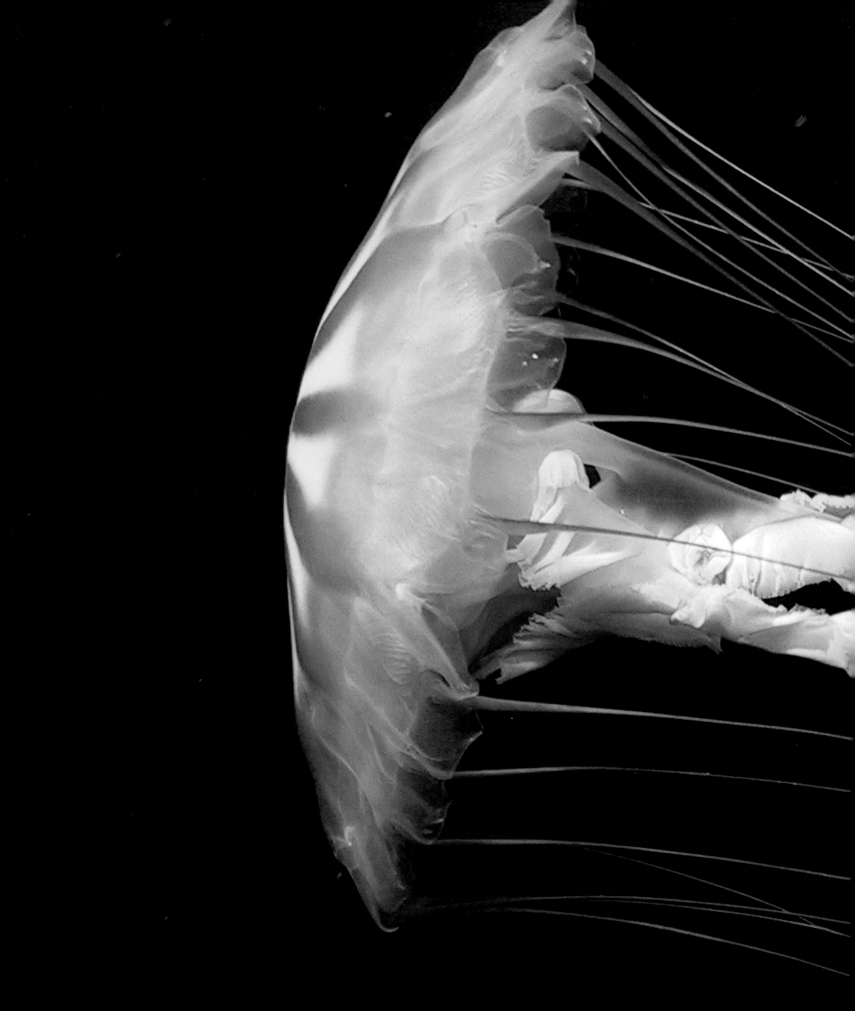

PART TWO

WHAT LIVES
IN THE OCEAN?

CHAPTER THREE

Expanding the Use of Technology

*We have studied everything from the small plants floating in the surface
waters, to the huge whales cruising the oceans, including the birds flying
over the waves, the fish swimming in the water column, and all sorts of
invertebrates swimming and floating around in the sea. We have also
examined the animals living in, on, and just above the ocean floor, from
the depths of the valleys to the tops of the seamounts of the Mid-Atlantic
Ridge. To accomplish this, we have been using high technology ranging
from satellites traveling in circumpolar orbits around the earth to highly
specialized equipment operating at the extreme depths of the seafloor.
It is with excitement that I wait for more such events, each bringing
to shore new knowledge on the mysteries of the ocean.*

— Birkir Bardarson, Pelagic Ecology Research Group,
University of St. Andrews, on the mar-eco cruise to the
Mid-Atlantic Ridge, August 16, 2007

Pages 74–75: A jellyfish (Chrysaora
melanaster) *moves through the water
in the high Arctic Ocean's Canada
Basin, an area surveyed as part of the
Census of Marine Life.*

*Opposite: A new species of octopus,
from the genus* Benthoctopus, *is
attached to an arm of the submersible
Alvin. It was collected in the Gulf
of Mexico.*

Travel and research in the ocean realms is like travel and research in
space. Both require complex technology, innovative ways of getting to
and from extreme environments, and, once there, the courage to investigate previously unexplored areas. Like space exploration, the Census of
Marine Life would simply not be possible without a vast array of technology, sophisticated gear and collaboration among scientists and engi-
neers who are willing to push the envelope to find new ways of seeing,
sampling and increasing understanding of life in the global ocean.

Census scientists are using nearly every trick in the book to accomplish the tremendous task of sampling the global ocean from top to

Space walk or deep dive? A deep-ocean researcher illustrates the parallels between space travel and work in the deep sea. Without the protection of this special type of submersible, called an atmospheric diving suit, this researcher's work would be impossible.

bottom and pole to pole, using tried and true tools of the trade and creating new technologies where none exist for the task at hand. Improved technologies are also allowing Census scientists to gain new understanding of previously studied areas. Ascertaining what is known, unknown and likely unknowable about marine life could not be achieved without the human ingenuity, creativity and stamina behind the discoveries being made by the Census of Marine Life.

REACHING THE RESEARCH SITES

The first-hand observation and sampling that are the staple of Census research require scientists to journey deep into the heart of the ocean. To make this journey, Census scientists depend upon several different kinds

The RV Polarstern, *an icebreaking research vessel, deploys research equipment onto the Antarctic ice. This unique vessel allows research to be conducted in areas that were previously unreachable.*

of vehicles. Large and small research vessels (RVs) are their commuter trains. Anything might be used to reach a marine study site, from small boats for near-shore work to very large ships capable of crossing the vast oceans and spending several months at sea. Providing a mobile platform for marine research, RVs typically carry a wide variety of sampling and surveying equipment; most have laboratory space on board so that researchers can begin to analyze the material they collect during the cruise. Some of the more advanced ships have special diesel-electric engines to minimize noise that may scare away fish and marine mammals. The Census of Marine Life uses a variety of RVs, including icebreakers that specialize in getting researchers into unique frozen marine habitats.

The end of the journey aboard an RV to a research site is just the beginning of the work of viewing and sampling marine life in its natural settings. To reach the depths of some ocean realms, scientists often use manned submersibles carried on the RVs. These underwater vessels are compact and self-contained but are dependent on their surface support vessels. Unlike the more well-known military submarines, research submersibles usually have a limited power supply and life-support capacity; they are designed for short-duration dives for the express purpose of collecting scientific data and samples. However, in some cases military submarines have been converted for oceanographic research. Some famous and well-used manned submersibles include the *Alvin* and *Johnson Sealink* from the United States and Russia's *Mir*. The Census also relies heavily on France's *Nautile*.

Working in a manned submersible requires patience and an ability to remain in very tight quarters for long periods of time. There is no room to stand in these vessels, so the many hours it takes to dive to a

The front of the Russian submersible Mir *is clustered with a variety of collecting and observing instruments.*

Above: The English ROV (remotely operated vehicle) Isis is being launched off the back of a research vessel.

Left: Pilots and scientists direct the ROV Isis from the control room of a research vessel.

research site and then return to the surface make for a very long work day! Considering the vastness of the deep sea and the extremely small field of view from a submersible, some of the known discoveries are nothing short of a miracle. Despite these disadvantages, groundbreaking accomplishments such as the discovery of hydrothermal vents in the late 1970s were achieved by ocean scientists tightly packed into submersibles, peering out of small round windows into the dimly lit abyss.

Manned submersibles, although important to ocean science research, are usually not capable of reaching the deepest ocean regions. They are also costly to operate and lack the versatility and endurance of unmanned remotely operated vehicles (ROVs), which give scientists an opportunity to study and collect organisms from greater depths without the risk to human life and with less expense, time and effort. ROVs are becoming one of the primary tools for studying the biodiversity of the deepest oceanic ecosystems, and are a key technology in Census research. They are linked to a surface support RV via an umbilical cord through which scientists control the ROV's underwater activity. This umbilical cord also supplies virtually limitless power and conveys data signals and video feed from the ROV's sensor array to the control station aboard ship.

Like remotely operated vehicles, autonomous underwater vehicles (AUVs) are useful unmanned exploration vessels. Like their ROV counterparts, they are capable of going deeper, for longer, and with less cost and effort than manned submersibles. Being autonomous, they are not tethered to the ship, and have the added benefit of being able to act and move on their own, collecting samples or data without requiring constant direct input from a surface-based RV control station. AUV operations can be controlled either by commands from an internal computer or by preset parameter files loaded into the vessel's control system before the mission starts. AUVs can carry camera equipment and oceanographic sensor packages that enable scientists to assess large areas of ocean space without having to actively pilot the vehicle, saving time and allowing other work to be carried on simultaneously.

A third type of unmanned vessel used by Census scientists is the deep-towed vehicle (DTV), which is pulled behind a research vessel as it traverses the ocean. DTVs are simpler than ROVs and AUVs, but they are useful as platforms for a variety of oceanographic instruments that measure biological, chemical and physical aspects of the ocean. There are many different types of towed vehicles, such as the moving vessel profiler (MVP), which can house a video plankton counter or similar device.

A Klein 3000 digital dual-frequency side-scan sonar is used for mapping surveys.

One advantage of using DTVs is that they can be equipped with external sensors to record data such as temperature and current speed. The DTV *Bridget,* used by the Census field project studying hydrothermal vents, for example, moves up and down near the ocean floor like a yo-yo in order to study the plumes of mineral- and chemical-laden water associated with hydrothermal vents.

USING SOUND TO "SEE" UNDER WATER

To achieve the Census's goal of quantitative estimates of species distributions and densities in many different realms and habitats, specimens must be observed and collected and data must be recorded about the environments in which they live. This is accomplished by direct sampling and by taking acoustic, chemical and optical measurements. The vastness of the ocean realms has prompted development of various innovative technologies to reach these goals.

Side-scan sonar is a type of acoustic technology that scientists use to "see" in the ocean; it is well-established as a technique for mapping the ocean floor and tracking schools of fish. Pulses of sound are projected by a ship or a device towed by a ship. These sound waves bounce off objects – whether they are living things or physical features of the ocean floor –

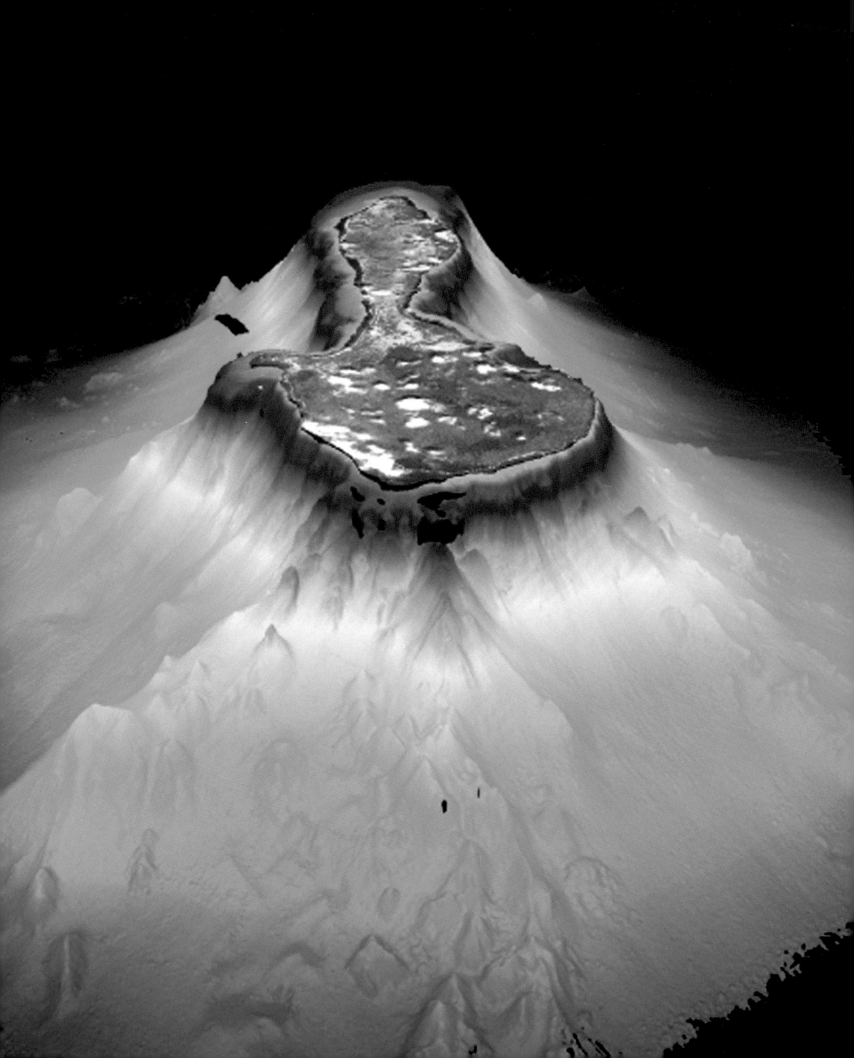

and are reflected back to the ship, where instruments convert them into images. The use of sonar is not confined to humans: biologists have learned that many animal species, including dolphins and bats, use echolocation – a type of natural sonar – to locate prey or to navigate.

Census scientists are using more sophisticated sonar systems called standard and multi-frequency echo sounders, or multibeam sonar, to estimate the size of plankton and fish populations. Echo sounders are also being used for species identification. This acoustic technology increases the success of imaging animals underwater because fish species respond differently to different sound frequencies, and they produce different reflected acoustic signals. Multibeam sonar has met with remarkable success and holds great promise for the future. In one Census study off the coast of New Jersey, for example, Census scientists found a school of 20 million herring that was roughly the size of Manhattan Island!

ADVANCES IN OPTICAL TECHNOLOGY

In addition to acoustic technology, advanced optical technology provides a relatively inexpensive and effective way of sampling large portions of the ocean for pelagic (free-swimming) organisms. One such technology is the video plankton recorder (VPR), in which a towed box moves water past a video camera that records images either continuously or at predetermined times. VPRs can be set to different resolutions to record a variety of planktonic organisms as small as some macroscopic diatoms, and they are ideal for imaging larger zooplankton such as copepods and the larvae of many other marine animals. VPRs can be towed by RVs or by commercial cargo ships traversing large areas of the ocean. Their capabilities are being steadily expanded by the Census project studying the Gulf of Maine.

Much is still unknown about the organisms that live at extreme ocean depths, but another type of technology used by Census scientists may help change this. Autonomous lander vehicles (ALVs) are designed to record time-lapse photographic images of marine life on the ocean floor down to depths of 6,000 meters (3.75 miles). ALV technology is proving a valuable tool for gaining understanding of the distribution, abundance and lifestyle of deep-sea benthic (ocean-floor) organisms. The basic ALV is a metal frame that supports a host of scientific instruments designed to measure physical properties such as conductivity, temperature, depth and current speed. Outfitted with high-resolution photographic equipment, these vehicles can autonomously record time-lapse images over a

Opposite: This three-dimensional visualization of Pagan Island, part of the Mariana Archipelago in the Pacific Ocean, was created from multibeam sonar data collected in 2007. It indicates a long shelf extending from the southern shore of the island, while other areas slope steeply down to 700 meters (2,300 feet) or deeper and show evidence of mass wasting and erosion.

period lasting anywhere from days to months. All landers are positively buoyant; when they have completed their tasks, weights can be released by an acoustic command from an RV and the ALV then rises to the ocean surface for recovery. This technology has been extensively used by the Census project studying the Mid-Atlantic Ridge, with great results.

COLLECTING SPECIMENS

The technologies described before help scientists observe and count organisms, but in some cases researchers must physically collect specimens. For instance, when they observe a previously unknown organism, examples of that species must be collected so that it can be officially identified or named. When scientists are unsure of what they may find at a particular site, before they invest time and expense in deploying more sophisticated research equipment, they may choose to sample the area, using various collection methods to determine whether it warrants closer study.

Trawling nets have long been used in oceanographic research, beginning with early studies of oceanic biodiversity, and Census projects use them extensively. Trawls are specialized large nets, akin to those used by fishers, in a variety of forms depending on the organisms of interest. Benthic trawls are used along the ocean floor, while pelagic trawls are used down to depths as great as 5,000 meters (3 miles). Some trawls sample in a series at different water depths to study the movement of organisms in the water column. Plankton nets are modified trawls used to collect intact planktonic organisms of nearly any size. Towed by an RV, plankton nets have a long funnel shape that terminates in a collection cylinder called a cod-end.

During a research cruise in 2006 on the RV *Ron Brown* to explore diversity in the deep Sargasso Sea, the Census of Marine Zooplankton field project successfully used three multiple opening/closing net and environmental sensing systems (MOCNESS) to trawl for zooplankton in deeper water than previously explored for these critters – 5,000 meters (3 miles). These newly designed trawl nets were fabricated from very fine (335 microns) nylon mesh, and the collected samples yielded a wealth of biodiversity from the deep water, including 13 species of cephalopods. Of these, three were octopods (*Cirrothauma murrayi*, *Bolitaena pygmaea* and *Tremoctopus violaceus*), one was a vampyromorph (*Vampyroteuthis*

This image of a mora fish (Mora moro) *was captured in water 1,000 meters (3,300 feet) deep in the northeast Atlantic by a camera attached to a ROBIO (RObust BIOdiversity) lander.*

infernalis) and the remaining nine were squids belonging to at least five major groups (bathyteuthids, chiroteuthids, cranchids, histioteuthids and enoploteuthids). The new 335-micron mesh net yielded specimens that were in immaculate condition, greatly simplifying taxonomic analyses.

This species from the squid genus Histeoteuthis *was discovered during a research expedition in deep waters of the Sargasso Sea by Census of Marine Zooplankton scientists.*

A trawl is lowered out the back of a research vessel during a scientific cruise sampling the mid-water depths of Astoria Canyon, 16 kilometers (10 miles) offshore from the mouth of the Columbia River.

A large variety of organisms can be caught in the net of a mid-water trawl.

Plankton nets are deployed to collect near-surface plankton. The net is about 2 meters (6.5 feet) long and has a mesh size of 236 microns (0.25 mm/0.01 inch).

Though trawls are useful in the study of marine biodiversity, they have their limitations. Many animals are very good at avoiding capture in the nets, and other species can be easily damaged or destroyed in the process, especially those from great depths and soft-bodied creatures such as jellyfish. Therefore trawls are often used in combination with other research tools such as video plankton recorders, acoustic technologies and larger imaging equipment.

When scientists want to sample organisms that live on or just below the surface of the ocean floor, they commonly use a benthic grab. Benthic grabs literally take a bite out of the seafloor. Various sizes and approaches work for different organisms and sediment types, but the aim is the same: to bring to the surface a complete sample of both the sediment and the organisms found in it. Using these techniques, scientists can describe the species found and make estimates of their abundance while preventing some of the injuries to delicate animals caused by other methods.

To collect and study organisms from the deep seafloor, scientists must also use specialized collection tools mounted on several types of submersible vehicles. Some, called suction samplers or "slurp guns," act like big vacuums that suck up small burrowing organisms from bottom sediments or free-swimming animals as they move through the water column. Other tools are specialized for grabbing intact clumps of animals attached to the bottom. And many submersibles, whether ROVs or manned, have mechanical arms that can reach out and grab a single sample of a species or a piece of the seafloor, ensuring that collected specimens remain in good shape – often alive and healthy – for better study. Many of the unique creatures that have been collected from the deepest regions of the oceans have been gathered using these tools.

STUDYING THE MOVEMENTS OF MARINE LIFE

Scientists collect organisms at a given place and point in time. If they want to know more about the collected animal's behavior, such as whether it migrates or whether it moves up and down in the water column, researchers will often locate them by using specialized intensive sampling technologies.

A large box corer is being hauled on deck after use. The standard box corer is designed to take undisturbed samples from the top of the seafloor; it is suitable for almost every type of sediment. It can penetrate the seafloor to a maximum of 50 centimeters (20 inches) using just its own weight. If necessary, the driving force can be adjusted by adding or removing lead weights, allowing for deeper penetration.

A steelhead salmon smolt is being readied to receive an acoustic tag.

Census research has advanced acoustic tag technology. Acoustic tags attached to animals use an audio signal to transmit information about the tagged animal as well as data on its depth, the water temperature and the amount of light in the surrounding water. This signal is picked up by either a mobile hydrophone or a series of receivers placed permanently under water to record the tagged animals when they travel within range.

There are several types of tagging technology. Some tags use tiny computers that record and store data such as the temperature, salinity and depth of the water where the tagged animals swim. Other tags may keep track of important clues about migration patterns, such as temperature – both inside the animal and outside in the water, salinity, depth and light availability.

Tagging technology has become more sophisticated because of work done in the Census program. Pop-up satellite archival tags (PSATs), which combine two types of technology to overcome the problems of tag recovery, are an example of how innovation is being used to resolve research problems. The basic tag collects the same data as a normal

Census scientists implant a satellite tracking beacon in a bluefin tuna. This beacon will allow researchers to record the movements of this highly migratory species, giving insight into its global patterns of distribution.

THERMAL IMAGING

An innovative technology that allows Census scientists to see a large area of the ocean at one time involves using satellites orbiting Earth. Some satellites travel in sync with Earth's spin, and therefore are fixed above one location on the planet's surface in a geostationary orbit. Others are in different types of orbits, such as from pole to pole, and can take global snapshots as they pass over different parts of Earth.

Satellites have proven to be invaluable tools for research; they are used in one way or another by almost every Census project. Satellite remote sensing is a technology used to determine different aspects of ocean conditions such as water temperature, chlorophyll levels (which indicate phytoplankton abundance) and oceanic currents. Satellites can even be used to track animals that have been tagged with beacons that transmit a variety of information. The Census project Tagging of Pacific Pelagics relies on advances in this satellite technology.

Above: This thermal image, taken by a geostationary satellite positioned over the western Atlantic, shows the Gulf Stream and northeastern coast of the United States. Several large Gulf Stream warm core rings are visible, as are the higher-productivity areas near Chesapeake and Delaware Bays. To the northeast, part of the Grand Banks region near Nova Scotia is visible. Despite the high productivity of this region, overfishing caused a total collapse of the Grand Banks cod fishery in the early 1990s.

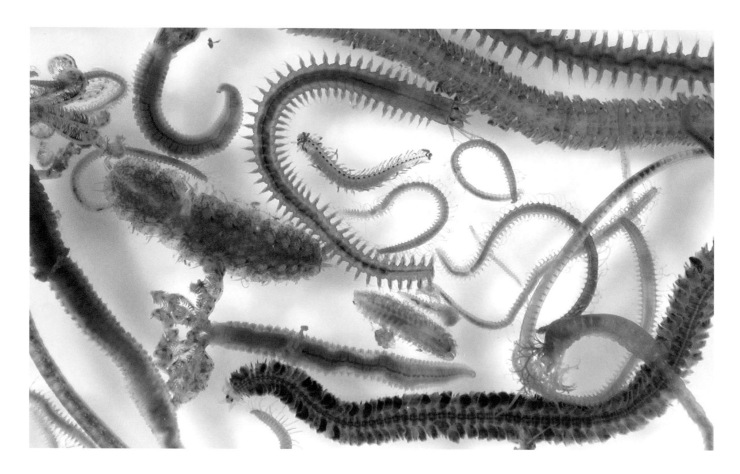

The diversity of marine species shown here demonstrates the challenges of taxonomic identification.

archival tag, recording measurements such as temperature, salinity, depth and light-based location, but it can also transmit its recorded information to an orbiting satellite, which relays the information to researchers. At a preset time, a battery is turned on that triggers transmission of the signal to the satellite. This battery activation also dissolves the tag attachment, allowing the tag to float to the surface, where it broadcasts its data and can possibly be recovered. While more expensive than other tags, PSATs have proven effective for studying the movements of large animals, such as sharks, that are not often possible to recapture.

Smart position and temperature (SPOT) tags use some of the most advanced technology employed by Census researchers today. Like the other types of tags, they record a variety of measurements such as temperature, salinity and depth. However, SPOT tags are powered by a very strong transmitter, so they can send their recorded data to satellites at regular intervals. These tags are designed primarily for use on animals that are commonly found on or near the ocean's surface – making regular broadcasts to a satellite possible – including dolphins, turtles, seals and any other animal that must regularly spend at least some time at the ocean surface. Recently these tags were successfully attached to the dorsal fins of sharks that swim at the surface.

IDENTIFYING THE COLLECTED SPECIES

The Census of Marine Life is collecting millions of organisms during its 10-year effort, and scientists anticipate discovering many thousands of different species. The process of accurately identifying so many organisms is a tremendous task. Traditionally, specialized scientists called taxonomists would use physical characteristics for species identification, but modern techniques also use an organism's genetic makeup for identification and classification. Combined with traditional approaches, genetic identification has given Census researchers an effective tool for cataloging the many marine organisms they discover.

In the traditional procedure for identifying organisms, the physical characteristics of a collected specimen are compared with those of a known species. Taxonomic keys and books describe the physical appearance – both external and internal – of millions of species, as well as what is known about their habitats and general biology. Census researchers study collected specimens, often through a microscope, to determine features such as the number of tentacles on a jellyfish or the length of the spines on a deep-sea anglerfish, and match what they find with existing species descriptions.

A new field of research has allowed for more accurate species identification – molecular techniques are being used to describe the genetic code of individual species. Comparison of these codes with genetic information extracted from collected specimens is a faster way to identify organisms than traditional methods. This method does not rely on an individual taxonomist's expertise or on the categorization of physical features, which are sometimes damaged or unclear. In addition, scientists can use molecular techniques to determine how different species are related to each other, allowing taxonomists to build a more thorough and accurate "tree of life" than has previously been possible.

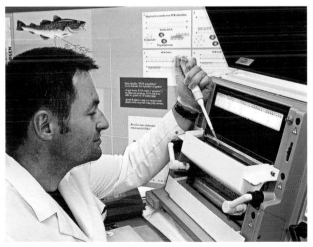

A scientist processes the DNA of an organism in order to determine its genetic makeup.

One important recent advance in the field of molecular techniques is the development of DNA barcoding, an approach that uses a small segment of an organism's DNA to identify its species name. During research cruises, Census scientists are using portable kits, and

This GIS image of the Gulf of Mexico was generated from numerous sources of information. The deeper the blue color, the deeper the water. The green denotes density of vegetation.

they were among the first to carry out barcoding aboard a ship on a rolling sea. This technology gives the researchers an advantage when they are trying to identify large numbers of collected organisms. DNA barcoding can isolate new species without having to go through the full process of describing and naming. This can be seen as both a benefit – for differentiating between closely related species or definitively determining taxonomic ties – and a drawback – because we can now identify new species faster than we can describe them, and some may never be formally described.

CENSUS DATA AVAILABLE TO THE WORLD

Census of Marine Life scientists are gathering a growing bank of data about marine life. They are analyzing and compiling this information, sharing it with a community of ocean scientists worldwide, and making it publicly accessible through an online database. Because all the data produced by Census projects are being made available, it is becoming possible to generate a comprehensive picture of marine biodiversity in the past, present and future.

Researchers are using both traditional and high-tech methods of visualizing how habitats and biodiversity change across geographic regions. These methods include standard mapping techniques, which can also be presented to show changes over time and space. The most commonly used approach, GIS (geographic information systems) mapping, uses computer technology to visually represent measurements of many different types of physical and biological characteristics for a specific geographic area. This technology is very helpful for examining population abundance, such as the density of plankton found in a bay, or physical characteristics of the environment, such as the temperature of seawater. Specialized computer programs can combine data for different characteristics on one map, with each characteristic represented by its own color. This creates a clear image of specific characteristics that can be compared to each other over the area in question.

Database managers are using computer technology to correlate and pool all the data gathered by the Census projects, such as species numbers and distribution, water temperature and nutrient availability. A key effort of the Census is an interactive online database called the Ocean Biogeographic Information System (OBIS), a Web-based provider of global, geographically referenced information on individual marine species. Users anywhere in the world can click on a map on their computer and bring up Census data on what lives in the ocean zone of interest. For example, OBIS enables the user to overlay images of the abundance and distribution of predatory populations on top of those of prey populations, shedding light on the inter-dynamics of the food web. To date, integrating such data throughout the water column has been difficult, but the shared standards and protocols of OBIS will make it easier, opening the door to improved understanding of the patterns and processes that govern marine life.

The effectiveness of research on marine biodiversity increases as researchers gain rapid access to data from previous studies. The OBIS

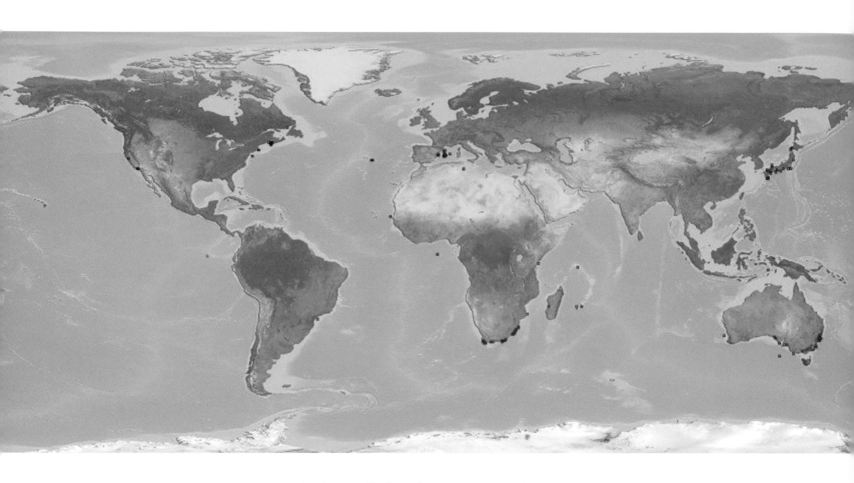

This map shows the known distribution (in red) of the great white shark (Carcharodon carcharias) around the world.

database will allow for production of long-term models and comprehensive pictures of marine life over time. It is a goal of the Census to have this next-generation information infrastructure in place and fully operational by 2010. The Census of Marine Life also promises to be the most important test bed for the technologies of observing marine life that will become part of the Global Ocean Observing System (GOOS), which governments have committed to building over the next two decades. This legacy of information is at the very heart of what makes the Census such a crucial component of our understanding of the diversity of marine life – past, present and future.

The many technologies employed by ocean researchers in general and by Census of Marine Life researchers specifically, and the fact that these technologies are continuously evolving, improving and being joined by new tools in their "bag of tricks," serve as a reminder that our knowledge of the ocean and its secrets is limited. While the primary goal of the Census is to create an organized set of data to serve as a baseline for future research and resource management, the 2010 conclusion of the

program will hardly represent the summit of the knowledge mountain. The simple fact that it requires all of the various scientific tools described in this chapter, plus a veritable multitude of others – including some not even invented yet – to accomplish a global census of the world's ocean sheds light on the complexity of the task of painting a clear picture of the inhabitants of the ocean in the past, present and future. Employing "every trick in the book" may make it possible to get a glimpse of what lives below the surface, an initial goal of the global census. As Jesse Ausubel, who originated the Census of Marine Life concept, has stated, "For marine life, the age of discovery is not over."

A great white shark (Carcharodon carcharias) *swimming with its mouth open; Gansbaai, South Africa.*

CHAPTER FOUR

Animals as Ocean Observers

We spotted a crabeater seal just after lunch. The seal team hopped into the Zodiac boat and chugged its way through large pieces of pancake ice. My heart started to beat a little faster the closer we got to the seal. Birgitte McDonald, a University of California Santa Cruz Ph.D. student, aimed the dart gun and fired. A direct hit of the tranquilizer went into the seal . . . my adrenaline going full speed. Finally, the nod was given. . . . Before I knew it, I was in an all-out wrestling match with a 600-pound seal on a block of ice in the Antarctic Ocean. It doesn't get any wilder than that! Once the seal was out, the seal team went into action, each person having a job of data collecting. Measurements, ultrasound, tissue biopsy, blood samples and weighing are the objectives of each capture. This was a great day! I do love being out here, and again, I have no idea what tomorrow will bring. All I can say is, "Bring it on."
— Mark Harris, a high school teacher from Layton, Utah, describing his experience tagging crabeater seals with Census researchers in the Antarctic

Opposite: Census researchers crowd onto Antarctic ice floes to tag seals with the newest generation of biologging technology. These tags will relay information about water temperature, salinity, swim speed and location to scientists, who use this data to study the seals and the ocean environment in which they live.

Recruiting animals as ocean observers, as Mark Harris describes above, is becoming an increasingly common method used by Census researchers in their quest to understand the ocean and its vast mysteries. Animal observers and the technologies associated with them are helping to fill in large holes in our understanding of the ocean. Areas of ocean space that were formerly too remote, too costly to reach or too dangerous for scientists to study are now becoming more accessible through the use of animal-borne instrumentation. These new technologies allow Census researchers to collect environmental data as the animal goes

The bluefin tuna, a highly migratory pelagicpredator, is one of the target species tagged by the Census field project Tagging of Pacific Predators (TOPP).

about its daily life in the ocean, and they also provide a window into the animal's eating, breeding, sleeping and other behaviors.

Two Census field projects are tagging marine animals to gain insight into the species' ranges, distribution and migration behaviors. By tagging more than 20 marine species, from North Pacific salmon to southern elephant seals in the Antarctic, Census researchers are pushing the envelope of tagging technology, making them among the forerunners in its development and use in marine science research.

DEVELOPMENTS IN TAGGING

The concept of tagging marine animals to gather oceanographic information is not new. Records of the practice go back to the 1930s, when Per Scholander used a basic mechanical depth gauge attached to a fin whale to measure its dive depth. Since that time, however, quantum leaps in technology have advanced the field of biologging, the practice of logging and relaying physical and biological data using tags attached to animals. Modern biologging is now providing environmental and animal behavioral data beyond Scholander's wildest imagination.

In their early days, animal-borne tags were used to collect information about an animal's environment, with the primary aim of furthering

knowledge in the fields of marine mammal physiology and behavior. The biologging systems of the 1960s and '70s were strictly mechanical, and included kitchen timers, depth gauges and paper scrolls among other readily available components. Early "tags" were primarily time/depth recorders (TDRs) that logged the dive time and depth of marine mammals such as elephant seals. The main limiting factor in these early systems was the challenge of recovering them for data retrieval: researchers had to be reasonably sure that they would reencounter the tagged animal. Those early TDRs were also cumbersome and heavy, which further limited their use to larger animals that would be more tolerant of the size and weight of the units.

Various improvements were made to TDR technology throughout the late 1970s and '80s. Some key advances were miniaturization of components, replacement of paper scrolls with film, and extension of run time — how long the tag can track and record data. The true revolution in tag technology, however, came with the application of microchips and integrated circuit electronic components, which caused a virtual explosion in the use of tagging technology for a wide range of ocean science research.

A crabeater seal surfaces through a break in the Antarctic ice.

A conductivity/temperature/depth (CTD) tag, a new generation of satellite-relayed data logger developed by the Sea Mammal Research Unit at Scotland's University of St. Andrew's, is the standard technology used for many tagging applications. In addition to its use with seals in the Antarctic, the CTD tag lends itself well to tagging other animals that spend significant time at the ocean surface, such as this leatherback turtle.

Electronic tags also allowed scientists to integrate a variety of auxiliary sensors into the tag complex. This has enabled Census tags to provide information on water depth, salinity and temperature as well as the animal's swim speed, profoundly increasing the richness of data that can be collected by these animal-borne instrument packages.

Another tagging innovation, the development of Argos satellite telemetry instrumentation, has opened another door. This technology for worldwide tracking and environmental monitoring allows researchers to track the movements of tagged animals, then cross-reference location information with data over time from the depth, salinity and temperature sensors. In essence, the data provide Census scientists with an oceanographic map of the area in which the animal moves. The age of animal research assistants in oceanographic science has indeed become reality.

Current tagging technology takes many forms. The specific technologies that apply to oceanographic data collection typically provide, at a minimum, location information and one or more types of environmental data. Census scientists are using a variety of these technologies in their research and have made some interesting discoveries.

The pelagic ocean sunfish (Mola mola) *is one of more than 20 marine species being tagged and tracked by Census scientists. This* Mola *carries a pop-up archival tag attached at the base of its dorsal fin.*

SATELLITE TAGS

A smart position and temperature (SPOT) tag broadcasts bursts of information to a satellite every time its antenna breaks the surface of the water. As the name suggests, these tags record an animal's position and the water temperature, as well as its speed and the water pressure (which indicates depth). Because this tag needs to break the surface periodically to send data to a receiving satellite, it is best used on animals that spend at least a portion of their lives at or very near the ocean surface.

Census researchers use SPOT tags to track salmon sharks along the Pacific coast of North America, for example, and have uncovered some surprising information about the range and migrations of these fish. Contrary to expectations, these warm-blooded sharks remain in the Far North during the winter, spending the coldest months of the year feeding in the freezing waters around Alaska. Because these waters are often icebound for much of the year, this aspect of salmon shark ecology would likely have gone unnoticed without the tagging research conducted by the Census. In addition, some salmon sharks surprised researchers by migrating as far south as the subtropical waters around Hawaii. Well outside the expected range for salmon sharks, these individuals have caused Census scientists to rethink the ecology and distribution of the species. Their long-range migrations, which include potential transboundary issues and extensions in their range, mean that management of these sharks, their prey species and the waters they frequent could be a bigger challenge than previously thought.

Researchers placing a smart position and temperature (SPOT) tag on the dorsal fin of a salmon shark.

Satellite-relayed data loggers (SRDLs) are sophisticated self-contained units embedded with various sensors that are attached to animals for periods ranging from months to years. Basic units used by Census scientists measure depth, temperature, salinity and swim speed. In addition, by communicating data via the Argos satellite array, the tags also record location information. A more recent innovation in SRDL technology has also made it possible to broadcast data via cellular phone frequencies. This allows very simple and cost-effective transfer of data in areas where cellphone network coverage overlaps with research areas.

SRDLs are the standard biologging technology that Census researchers are using to track seals in the North Pacific and the Antarctic. These tags have provided information about seals' annual migrations,

This male elephant seal has two tags on its head. The main one is a satellite tag that allows researchers to track the animal across the ocean. The smaller tag on top is a radio transmitter tag that allows the instrument to be recovered when the animal is on the beach.

which scientists cross-referenced with environmental data to provide valuable information about the seals' feeding preferences. This new knowledge posed questions about other poorly understood behavior. For instance, if a southern elephant seal remains where the sea surface temperature makes the area prone to icing over, one might wonder why it doesn't simply move to warmer waters. Using temperature and salinity data collected by the animals' SRDL tags, Census scientists discovered that environmental conditions near the lower extent of the elephant seals' dives make a perfect habitat for their favorite prey. This helped the researchers understand the previously unexplained, rather mysterious behavior of these seals. As an added bonus, Census researchers were able to generate oceanographic charts of the environmental conditions in these areas.

Archival Tags

A pop-up archival tag can be attached to almost any animal that can bear its weight, no matter how much time is spent at the surface, because the tags are programmed to break away after a set period and float to the surface. When they reach the surface of the water, the tags broadcast their data to satellites until their batteries die. The data provided by these tags include water temperature, depth profiles and ambient light levels. Currently they are also able to estimate latitude and longitude down to one degree.

Census researchers have used pop-up archival tags to successfully track sharks, tuna and ocean sunfish. By using the tags, scientists are learning more and more about the environment that these animals prefer. For example, Census researchers discovered a specific area in the Pacific Ocean where great white sharks tend to congregate during certain parts of the year. Until now, little was known about white sharks in the pelagic (open ocean) environment. The sheer vastness of this virtually barren environment – sometimes referred to as the "blue desert" – has made the study of its inhabitants difficult at best. However, attaching pop-up archival tags to white sharks has opened a door to their pelagic migrations and the conditions of the ocean they frequent.

The scientists discovered that, every year, tagged sharks congregate in an area of the North Pacific they have dubbed the "White Shark Café." Sharks from different home ranges tend to cluster in this area, spending a significant amount of time there and repeatedly diving to great depths. While scientists cannot yet determine what role this migration and diving behavior may play in the sharks' life cycle or ecology, environmental data retrieved from their tags is helping in Census researchers' attempts to explain this mysterious behavior.

Tracking Animals by Sound

Archival acoustic tags are a new generation of acoustic tag technology that uses fixed listening (receiving) arrays to communicate an animal's position. The tags contain archival sensor packages that store depth and water temperature data. When a tagged animal swims past a listening array, the tag broadcasts the archived data to a receiver in the array. This

Yurok tribe member and biologist Barry McCovey Jr. (left) and Yurok Tribal Fisheries staff member Scott Turo prepare to release a green sturgeon. Tagged by Census researchers in California, the fish was later detected off British Columbia, surprising scientists, who had never encountered this species so far north.

technology has proven to be a cost-effective method for collecting oceanographic data by using animals that frequent nearshore habitats, such as Pacific salmon and green sturgeon.

Preliminary results have shown that salmon smolt survival follows patterns that are contrary to what was formerly accepted. Fisheries biologists have long thought that salmon mortality is higher on rivers that have been dammed than on those that flow unimpeded. Census researchers are producing results that prove otherwise. Tags have been used to demonstrate that the survival rates of salmon smolt on the Columbia River, which is a heavily dammed system, are equal to or higher than those of smolt on the Fraser River, a naturally flowing river system. This suggests that the hardships for young salmon are no worse on dam-affected river systems, and that the true trials of life for these fish begin after they enter the saltwater environment. Differences in mortality from one river system to the next could thus have more to do with their associated local saltwater environments than with conditions on the rivers themselves. Because this information contradicts accepted theories of salmon ecology, Census work is likely to have a significant impact on how these fisheries are managed in the future.

Acoustic tags are surgically implanted in the abdominal cavity of target fish. These tags collect data and broadcast it to receivers deployed in an array along the Pacific coast of North America.

Acoustic receivers, such as the ones shown here ready for deployment, pick up the signals emitted by acoustic tags. The receivers are anchored to the seabed and form a "listening curtain" that receives data from tagged animals as they pass in or out of the curtained area.

FACING EXTREMES

While many different animals have been the subjects of Census tagging research, when it comes to collection of oceanographic data, the main players tend to be larger animals that migrate across great distances or have patterns of longer-range movements. Marine mammals such as seals, whales and sea lions; pelagic fish such as tuna, ocean sunfish and sharks; anadromous fish (which swim up rivers from the sea to spawn) such as salmon; sea turtles; various marine bird species; and certain squid species all exhibit the type of movements that make them ideal candidates for use as ocean observers.

Advancements in our knowledge of the ocean have come from Census scientists' recruiting of these highly migratory animal observers to collect oceanographic data in less sampled areas such as the polar oceans, the pelagic environment and the deep sea. Tagging technology has reduced dependence on manned submersibles, remotely operated vehicles and autonomous underwater vehicles for collecting data. Tagging marine mammals that make repetitive dives means that data can be collected more safely, with fewer risks for the human researchers.

Census researchers brave the Antarctic environment to tag seals that will relay both behavioral and oceanographic data.

SEAL TAGGING

Furiously working on an ice floe barely larger than their Zodiac, researchers in the Antarctic hurry to finish tagging a crabeater seal before the last bit of their six hours of daily light disappears.

Faced with bitter cold and howling winds and working on ice floes barely large enough to support their team, marine mammal expert and Census scientist Dan Costa and his crew have been tagging elephant seals and crabeater seals in the Antarctic for several years. Their work exemplifies the cutting-edge research of the Census of Marine Life and its field projects. By using the latest biologging technology, they have been able to shed light on the foraging behavior of these seals and also to further our oceanographic knowledge of the Southern Ocean. According to Dan, "These tags transmit information on the seal's location and its diving behavior, as well as data on the temperature and salinity of the water as it dives. Such data are important to understand why the seals go where they go, as well as to understand the basic physics of the ocean. In fact, such data are quite valuable to physical oceanographers, as they can use them to model oceanic currents."

Above: A tagged seal snorts a final goodbye to researchers before embarking on its data-collection mission.

Right: One obstacle that Census researchers have to face when tagging seals is the possibility of a bite. The teeth of a crabeater seal, while adapted to straining krill from seawater, remind scientists that these massive animals are not to be underestimated.

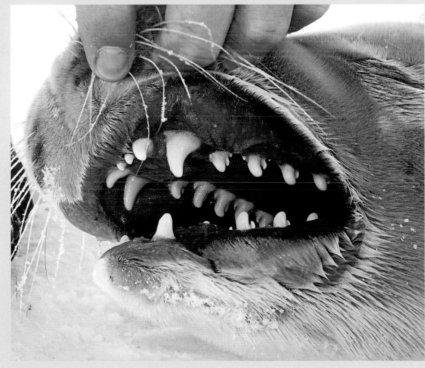

An Antarctic fur seal mother and pup represent current and future generations of potential animal observers. With the help of these research assistants, scientists may be able to prevent the loss of their polar habitat to global climate change.

Fighting through 20 centimeters (8 inches) of pack ice in a small inflatable boat, wrestling temperamental seals on floating chunks of ice and getting blown about by the weather are all in a day's work for these intrepid scientists. Mark Harris, a member of the team, recorded in his research journal: "Mother Nature has turned on us. The temperature was 18°F [–8°C] with 50 mph [80 kph] winds, snowing with a wind chill of minus 48°F [–44°C]. Not a good day to catch seals; it was decided early on that there would be no operations today."

Although the scientists may spend a month or more at a time in these icy waters, they might only manage to tag a dozen or so seals. However, the team's fortitude in steadily tagging as many animals as they safely could during their 2006 expedition has paid off. The oceanographic data collected by

Costa's seals have been validated against traditional oceanographic methods and have been found to be both accurate and useful. This approach has effectively created a new tool, as accurate as traditional methods, for gathering oceanographic data.

Much of Costa's data describe ocean space underneath the ice, an area that has historically been undersampled because of logistical and technological limitations. Seal-collected oceanographic data promise Census researchers an opportunity to look into this "blind spot" that may prove invaluable to scientists studying global climate change. Antarctic bottom water is a key component of oceanic circulation, which drives the global climate, but measuring its characteristics is difficult. Data collected by animal observers may help in understanding this environment and the mechanisms by which it is driven and by which it may change.

"Seal Team 1" poses with its latest recruit in its mission to collect data about the undersampled Antarctic ocean. "Seal tagging and getting on an ice floe below the Antarctic Circle in the dead of winter down here — that is about as wild as it gets!" says Mark Harris.

BLUEFIN TUNA: SEAFOOD, SATELLITE TAGS AND THE CENSUS

The majesty of the bluefin tuna has awed humans throughout our history. Their power, grace and predatory prowess have propelled them to iconic status in many societies. However, their delicious flavor is contributing to their decline. Bluefin tuna, known in the Japanese market as kuromaguro, is in high demand worldwide and commands premium prices, even as stocks steadily wane. Hence, the effort put into catching this iconic seafood has reached huge proportions.

Experts are now warning that bluefin stocks cannot handle this pressure, and much work is being done to gather information about the species in an attempt to better manage it. Says Barbara Block, chief scientist of the Census of Marine Life Tagging of Pacific Predators (TOPP) program, "Their population is on the brink of collapse, and it has happened on our watch, in my lifetime. We have the science today to rebuild populations and prevent them from going the way of the cod."

Bluefin tuna is one of the species being intensively studied by Census researchers. Scientists using satellite tags to track bluefin migration and diving behavior have found that these giants of the sea will routinely cross entire ocean basins during their annual movements. Tuna that were tagged while swimming together in Ireland have been found more than 4,800 kilometers (3,000 miles) apart less than eight months later. Results are proving that traditional management strategies for these tuna – and the notion that ocean space has boundaries – don't take into account these vast migrations, making successful conservation of the stock unlikely. Bluefin tuna don't respect boundaries drawn by humans. Because of their global-scale migrations, fishing in European waters affects stocks in the Gulf of Mexico, and bluefin poaching in the Southern Ocean affects North Pacific stocks. Collaborative, cross-boundary management is emerging as a key necessity if tuna stocks are to recover and survive.

Tagging efforts by Census researchers and affiliated groups – such as the Tag-a-Giant Foundation, the Large Pelagics Research Center and the Pelagic Fisheries Conservation Program, to name just a few – are helping to improve management efforts by providing a clearer picture of how these iconic fish use ocean space. Increased understanding of their ocean habitat, coupled with oceanographic data collected by sophisticated electronic tags, is helping scientists to better comprehend how environmental conditions affect tuna feeding and spawning ecology. Ultimately, this knowledge could help prevent their demise.

A trio of bluefin tuna demonstrates the majesty of the species. Stocks of this highly sought-after seafood are waning. However, Census tagging has shed light on bluefin ecology and may provide valuable information for future management of this species.

Advances in tagging technology allow researchers to recruit highly migratory bluefin tuna, such as the one seen here being implanted with a tag, for the task of oceanographic data collection.

Oceanographic sampling in the polar regions has been a challenge since the early days of exploration. Polar oceanography requires specialized equipment, research vessels and people power and is subject to limited field seasons that are often encumbered by sea ice. It has thus been, at best, both expensive and difficult. Animal-borne sensor arrays have given research in the polar oceans a tremendous boost. Sampling is no longer encumbered by the limited capabilities of human scientists working during restricted research seasons. Animal observers are now being fitted with tags that allow them to sample and report on environmental conditions as they go about their lives, long after visiting scientists have departed the area.

Tagging has provided a similar boost to research in the pelagic ocean and the deep sea. Animal research assistants are opening windows of knowledge in these areas that have traditionally been undersampled because of cost and safety issues and the relative vastness of the environments. By tagging animals such as highly migratory tuna and sharks, Census researchers studying the pelagic ocean environment have uncovered surprising biological richness in certain areas. So-called biological hotspots in the Pacific Ocean – veritable oases in the vast open sea – have been revealed. Likewise, tagging deep-diving animals such as the elephant seal described earlier has helped reveal the richness of some deep-sea habitats and their relationship to ecosystems in shallower waters. Previously, the deep-sea environment had proven to be one of the most

An unlikely research assistant, this tagged juvenile salmon is one of many species that are helping scientists better understand the marine environment.

challenging ocean realms to study, but with deep-diving seals, whales and the like carrying instrument arrays into the abyss for Census researchers, the volume of data that can be collected is expanding every year.

It is possible to envision an armada of animal observers, tagged with the latest, most sophisticated biologging equipment, swimming the seas, living their lives and at the same time passively furthering our understanding of the ocean. Census research is helping to make this vision a reality. As technology advances, batteries get smaller and more powerful, and sensors get more accurate, more durable and less costly, Census research will benefit from greater precision and a clearer focus. While it is unlikely that animal observers will fully replace scientists in the study of the ocean environment, it appears increasingly likely that, in time, scientists will have a whole lot more help from the very creatures being studied.

THE LONGEST ANIMAL MIGRATION: THE SOOTY SHEARWATER

Because of advances in the miniaturization of tagging technology, animals as small as sooty shearwaters can be tagged and tracked, shedding light on their ecology and environment.

A 64,000-kilometer (40,000-mile) journey is quite a trip, but imagine traveling that distance as a small bird! Even though global air travel has made the world a smaller place, most people on this planet won't travel this distance in their entire lifetime. A small shorebird named the sooty shearwater, however, routinely covers this distance every year on its annual migration from New Zealand to the North Pacific and back again. Census researchers deployed tracking tags on 33 sooty shearwaters in New Zealand in early 2005. Later that year, when the shearwaters returned to New Zealand from their annual migration, 20 tags were recovered. The tale they told shocked researchers and substantially changed scientists' understanding of shearwater ecology.

Data recovered from the tags showed that sooty shearwaters traveled up to 880 kilometers (550 miles) a day on their migration from breeding grounds in New Zealand to North Pacific feeding sites in Japan, Alaska and California, and some individuals traveled up to 62,400 kilometers (39,000 miles) in a season! To date this is the longest-ever recorded animal migration. The tags, which also record water pressure to indicate depth, showed that, in addition to all this traveling, sooty shearwaters were diving to more than 60 meters (200 feet) in pursuit of fish, squid and krill.

Researchers hope that continued tracking of shearwaters will provide further insight into their migratory pursuit of an endless summer and help scientists forecast how global climate change could affect their populations. Also, since studies show that populations of sooty shearwaters are in decline, precise migration data can aid their conservation by identifying critical feeding and breeding areas across the entire Pacific Basin.

CHAPTER FIVE

Disappearing Ice Oceans

More than scientific curiosity drives this enterprise.
The planet's polar regions are uniquely vulnerable to the effects of global
climate change. They are reacting to the planet's gradual warming in
increasingly unexpected and dramatic ways.

– RON O'DOR, CO-SENIOR SCIENTIST, CENSUS OF MARINE LIFE

Dramatic change is occurring at the opposite poles of our planet. A climatic shift caused by global warming is leaving an indelible mark on the "ice oceans" and the creatures that inhabit them. The ice at the poles has always grown and shrunk with the seasons. However, as global average temperatures rise, the overall amount of ice at the poles is shrinking. Each year during the month of September, the amount of sea ice floating in the Arctic Ocean is typically at its lowest for that year. In 2007, however, the loss of Arctic sea ice in September set a modern-day record: the ice cover shrank to about 4.1 million square kilometers (1.6 million square miles) – 43 percent less than in 1979, when accurate satellite observations first began.

Similarly, the Antarctic ice sheet, which covers about 98 percent of the continent and has an average thickness of more than 1,600 meters (1 mile), is losing approximately 150 cubic kilometers (36 cubic miles) of ice every year, according to a recent study that used satellite observations.

Because of these dramatic, relatively rapid changes, there is a sense of urgency among scientists studying life in the ice oceans to learn as much as they can, as quickly as they can, before the regions are irreversibly

Opposite: Scientists aboard the RV Polarstern *captured this picture of a drifting plateau iceberg in a very calm sea in Antarctica. The iceberg was one of thousands seen by researchers during a 10-week expedition to the Weddell Sea in 2006–07.*

117

This sea-ice pressure ridge was photographed from a scuba diver's perspective during a 2005 expedition to the Canada Basin.

changed. Fortunately, a means of studying these regions more fully was already in the works with the designation of the fourth International Polar Year (IPY) by the International Council for Science and the World Meteorological Organization (the previous IPYs were in 1882–83, 1932–1933 and 1957–58). To allow for full and equal coverage of both the Arctic and the Antarctic, the fourth IPY (actually two years – from March 2007 to March 2009) brought together thousands of scientists from more than 60 nations to examine a wide range of physical, biological and social research topics. Two Census projects are playing an integral role in the IPY initiative; in the Antarctic region alone, the Census has gathered data from 18 expeditions during this two-year period.

Scientists are racing to study these areas before they are further affected by climate change so that they can establish a baseline against which future changes can be measured. Researchers are also seizing the

This satellite image shows the minimum average Arctic ice for 1979–81. Satellite observations first became available in 1979.

This image illustrates the difference in the minimum average Arctic ice for 2003–05.

Arctic explorers have to be prepared for cold, brutal conditions as they investigate previously inaccessible polar environments. This is a typical sea-ice station during a 2005 expedition to the Canada Basin in the Arctic Ocean. The equipment includes a generator for power, sleds for hauling the gear, an ice corer, and light and temperature sensors.

opportunity to discover what inhabits these regions before they are altered by climatically induced environmental changes.

"The tremendous ongoing changes in the Arctic make the effort to identify the diversity of life in the three major realms [sea ice, water column and seafloor] an urgent issue," explains Rolf Gradinger, who, with his colleagues Bodil Bluhm and Russ Hopcroft, is leading the Census effort in the Arctic Ocean from their base at the University of Alaska Fairbanks. The magnitude of predicted environmental change and its effects on marine life require long-term monitoring, crucial to which is the availability of baseline data. "Species-level information is essential to discussions of climate change, its expressions and effects," Gradinger adds.

Gradinger's colleague in the Southern Hemisphere agrees. "What we learned from this expedition [aboard the RV *Polarstern* from December 2007 to January 2008] is the tip of an iceberg, so to speak. Insights from this and upcoming International Polar Year voyages will shed light on how climate variations affect ice-affiliated species living in this region," explains Michael Stoddart of the Australian Antarctic Division, one of the leaders of the Census Antarctic research program.

Certainly these scientists have their work cut out for them. Conducting scientific investigations in these regions is cold, difficult and dangerous

Remotely operated vehicles (ROVs), such as the one shown here, offer scientists eyes and ears below the ice, as well as a means to collect samples for later observation and study. This ROV was used to study the Arctic's Canada Basin, a huge, largely ice-covered underwater hole some 3,800 meters (2.4 miles) deep.

– not to mention expensive. In fact, research in the Southern Ocean is the most expensive in the world: a 2005 Census icebreaker expedition to the Southern Ocean cost one euro (about US$1.40) per second! Polar exploration requires careful planning and the services of sophisticated ice-breaking ships to access the remote, frigid environments at the poles of the earth. In spite of the tremendous challenges, Census explorers have made, and are making, real inroads into increasing knowledge about what lives below the oceanic ice.

Since the polar regions have been underexplored because of their relative inaccessibility, the rate of discovery when these areas are sampled has been simply remarkable. New life-forms have been found on virtually every expedition to these remote regions. It is estimated that half of all species found below 3,000 meters (about 2 miles) in the global ocean are new to science; in isolated parts of the world, such as the Southern Ocean, the figure may be closer to 95 percent. In three expeditions to the Southern Ocean from 2004 to 2007, for example, more than 700 new species were discovered.

UNCOVERING HIDDEN OCEANS

In 2005, 24 scientists from four countries (the United States, Canada, Russia and China), including 11 Census researchers, boarded the U.S. Coast Guard cutter *Healy* for a 30-day journey to the planet's most

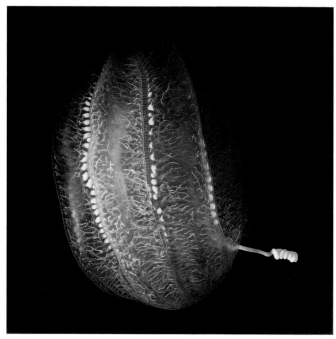

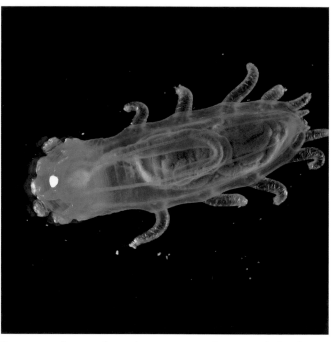

This comb jelly, an Aulacoctena *species, was collected by an ROV in the deep Arctic Canada Basin.*

Sea cucumbers such as this specimen dominated the fauna on the seafloor at several stations during the 2005 Hidden Ocean expedition.

This feather star was collected by an ROV during the Canada Basin Hidden Ocean expedition.

This High Arctic sea star was collected at the deep seafloor in the Canada Basin, using an ROV.

Sampling methods were chosen to minimize impact on the environment. One of these methods was scuba diving. Here divers prepare to go below the ice.

northern reaches. Their expedition during the brief polar summer revealed a surprising density and diversity of Arctic Ocean creatures.

During this voyage, unexpectedly high numbers and varieties of large Arctic jellies, squid, cod and other animals were found thriving in the extreme cold, sheltered for millennia under a lid of ice up to 20 meters (70 feet) thick. *Healy* returned to port from its "Hidden Ocean" expedition – funded by the U.S. National Oceanic and Atmospheric Administration – with thousands of specimens from the Chukchi and Beaufort Seas and the Canada Basin, a vast bowl walled by steep ridges and topped with ice.

The scientists, operating around the clock to maximize expensive ship time, sampled at 14 locations, 5 of them at depths of 3,300 meters (2 miles) and more. The explorers employed a suite of sampling tools that included a remotely operated vehicle (ROV), a benthic camera platform, under-ice cameras and scuba divers, complemented by pelagic nets, benthic box corers and an ice corer.

To explore the Arctic, one has to have both a sense of adventure and a sense of humor. Diver Elizabeth Siddon demonstrates both here.

Amid several thousands of specimens and images collected, scientists suspect they have found new species of jellyfish and benthic bristle worms, and the first squid and octopus ever found in the Arctic Ocean. It may be years, however, before the specimens are formally designated as new species; the peer-reviewed process to do so is exacting and detailed, requiring collaboration with expert taxonomists and comparisons with other similar species. In addition to potentially new species, scientists were also intrigued to discover two species of sand flea-like crustaceans, or amphipods, which, though familiar, had never before been found in an ice environment.

"Overall, the densities of animals were much higher than expected," says Census researcher Bodil Bluhm. "It now appears possible to confirm that the rich biodiversity surprising deep-sea explorers worldwide exists as well in deep Arctic waters, the most understudied area of the ocean world."

SURPRISES IN THE SOUTHERN OCEAN

The Southern Ocean is made up of the southern portions of the Atlantic, Pacific and Indian Oceans, but it is sometimes considered a separate ocean because it has lower average temperatures and salt concentration than the water bodies that contribute to it. It covers 35 million square kilometers (approximately 22 million square miles), or 10 percent of the world's ocean surface, and it is the planet's most mysterious and treacherous sea. For centuries sailors have called it the "Roaring 40s," a reference to the ceaseless winds that scream across Earth's barren southern latitudes.

The Southern Ocean is also influenced by the Antarctic Circumpolar Current, a powerful force flowing from west to east that moves 145 million cubic meters of water per second, or more than the combined flow of all the world's rivers. This intense current creates a physical churn in the Southern Ocean that scientists are investigating in an attempt to determine its role in the distribution of life in the global oceans and in continued climate change.

One Census expedition targeted an area of 10,000 square kilometers (4,000 square miles) of the Antarctic seabed that was made suddenly accessible as a result of climate change. A group of 52 marine scientists were the first to conduct a comprehensive biological survey of the area uncovered by the collapse of the Larsen A and B ice shelves in 1995 and 2002 respectively — the area opened up by the collapse is about the size of Jamaica.

Working aboard the German icebreaker RV *Polarstern*, operated by the Alfred Wegener Institute for Polar and Marine Research, the researchers spent 10 weeks investigating icy waters as deep as 850 meters (2,800 feet) off the Antarctic Peninsula. Their mission was threefold: to chart the environmental impact of history's largest known ice-shelf collapses, to find what endemic forms of marine life existed under Larsen

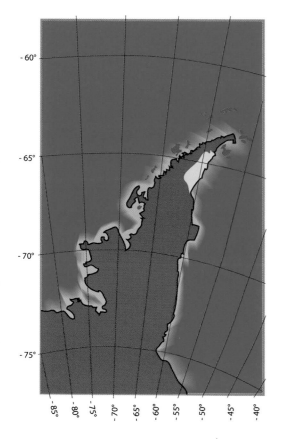

The area in yellow shows the approximate location of the Larsen A and B ice shelves before their collapse.

When large Antarctic glaciers reach the coast of the continent, they float, being less dense than seawater, and become part of an ice shelf, calving large blocks to form icebergs in the sea. Since 1974 a total of 13,500 square kilometers (5,400 square miles) of ice shelf has disintegrated off the Antarctic Peninsula, a phenomenon linked to regional temperature increases in the past 50 years. Scientists worry that similar breakups in other areas could lead to increased ice flow into the sea and cause sea levels to rise. Shown here is an evening view of the location of the former Larsen B ice shelf.

Part of the broken connection between the Larsen B ice shelf and the Antarctic Peninsula. The picture was taken during the RV Polarstern *expedition ANTXXIII/8 in the Weddell Sea, 2006–07.*

A and B, and to learn what new organisms may have opportunistically moved into the area after the collapses. Using sophisticated sampling and observation gear, experts on *Polarstern* gained a wealth of new insights and discovered several unfamiliar and potentially new creatures among an estimated 1,000 species collected.

"The breakup of these ice shelves opened up huge, near pristine portions of the ocean floor, sealed off from above for at least 5,000 years, and possibly up to 12,000 years in the case of Larsen B," explains Julian Gutt, a marine ecologist at Germany's Alfred Wegener Institute and chief scientist on the *Polarstern* expedition. "The collapse of the Larsen shelves may tell us about the impacts of climate-induced changes on marine biodiversity and the functioning of the ecosystem."

Until this expedition, scientists had glimpsed life under Antarctica's ice shelves only through drill holes. No longer constrained by the ice, the scientists aboard *Polarstern* were in the unique position of observing first-hand a marine ecosystem considered one of the least disturbed by humankind anywhere on the planet.

Sea cucumbers were abundant in the Larsen B area. Interestingly, here they all seem to be headed in the same direction.

Scientists made a number of interesting observations from their unique vantage point on the RV *Polarstern*. They found that the sediments of the Southern Ocean seafloor are extremely varied, ranging from bedrock to pure mud. As a result, the animals living on the seafloor, or epifauna, are also highly varied, though far less abundant in the Larsen A and B areas – perhaps only 1 percent as abundant as seabed animals found in the eastern part of the Antarctic's Weddell Sea. Despite this lower abundance of epifauna, scientists were intrigued to find abundant deep-sea lilies (members of a group called crinoids) and their relatives, sea cucumbers and sea urchins, in the relatively shallow waters of the Larsen zone. These species are more commonly found deeper than 2,000 meters (1.3 miles); they are able to adapt to life where resources are scarce – conditions similar to those under an ice shelf.

Apparent newcomers found colonizing the Larsen zone include fast-growing gelatinous sea squirts. The scientists found dense patches of sea squirts, and say they are likely to have been able to colonize the Larsen B area only after the ice shelf broke free in 2002. Very slow-growing animals called glass sponges were also discovered; their greatest densities were in the Larsen A area, where life-forms have had seven more years to recolonize than around Larsen B. The large number of juvenile forms of glass sponges observed probably indicates shifting species composition and abundance in the past 12 years. Census researchers also found a number of potentially new species, which are awaiting taxonomic analysis.

Above: These fast-growing ascidians, or sea squirts, were found at Larsen A, and could indicate a natural shift in biodiversity following the collapse of the ice shelves. The animal in the foreground has been colonized by two crustaceans and a brittle star.

Right: At Larsen A the expedition found large glass sponges, which are extremely slow-growing, which means they were likely there before the recent disintegration of the ice shelf.

In all, the *Polarstern* and its helicopter crews covered some 700 and 8,000 nautical miles (1,300 and 15,000 kilometers) respectively as they recorded the presence and behaviors of Southern Ocean marine mammals. Significant observations included minke whales close to the pack-ice edge and very rare beaked whale species near Elephant Island. "It was surprising how fast such a new habitat was used and colonized by minke whales in considerable densities," says specialist Meike Scheidat from the Research and Technology Centre Westcoast in Büsum, Germany. "They indicate that the ecosystem in the water column has changed considerably."

According to Census Antarctic project leader Michael Stoddart, a significant consequence of rising temperatures in the Antarctic Peninsula is the slow decrease of sea ice and of the planktonic algae that grow underneath. These algae feed krill, which are small shrimp-like creatures, and thus represent the lowest link in a marine food web that eventually sustains the iconic large Antarctic species – penguins, whales and seals. An adult blue whale eats about 4 million krill per day. "Algae are a source of abundant, high-quality winter food and are utterly central to the health of the whole ecosystem," says Stoddart. He adds that recent research by colleagues from the United Kingdom shows that krill stocks are decreasing significantly around the Antarctic Peninsula.

OPENING WINDOWS OF KNOWLEDGE

In 2008 another three Antarctic research vessels returned from the Southern Ocean. Like their earlier counterparts, they had collected a vast array of ocean life, including a number of previously unknown species, from the cold waters near the East Antarctic land mass.

Australia's RV *Aurora Australis* and its partner vessels *L'Astrolabe* (France) and *Umitaka Maru* (Japan) – all participants in the International Polar Year – demonstrate how collaboration not only expands the use of resources but also vastly enhances the knowledge and insight gained from these expeditions. The scientific teams aboard the French and Japanese ships examined the mid- and upper-ocean environments while the team aboard *Aurora Australis* focused on the ocean floor. Technology played a key role in what they learned.

The voyage leader on the RV *Aurora Australis*, Census researcher Martin Riddle of the Australian Antarctic Division, says that their expedition uncovered a remarkably rich, colorful and complex range of marine life-forms in this previously unknown environment. They used

Sadie Mills (left) and Niki Davey hold giant Macroptychaster *sea stars – up to 60 centimeters (2 feet) across – that were collected in Antarctic waters.*

video and digital still cameras mounted on the top beam of a trawl. "The material collected by a trawl can look a bit like a marinara mix when it comes on the deck, but the cameras allow you to see exactly what the animals were like in their undisturbed state," says Riddle. "In some places every inch of the seafloor is covered in life. In other places we can see deep scars and gouges where icebergs scour the seafloor as they pass by. Gigantism is very common in Antarctic waters – we have collected huge worms, giant crustaceans and sea spiders the size of dinner plates."

Martin Riddle suggests that these surveys establish a point of reference for monitoring the impact of environmental change in Antarctic waters. Of particular importance is the discovery of cold-water coral communities at a depth of about 800 meters (2,640 feet) in canyons leading off the continental shelf into deeper water. These calcium carbonate–based communities are likely to be among the first marine communities affected by ocean acidification, which is caused by rising levels of atmospheric carbon dioxide (CO_2).

Increased CO_2 in the oceans increases the amount of carbonic acid, which in turn makes calcium carbonate – used by many marine creatures

A giant sea star on the seabed of the Southern Ocean.

to construct shells and skeletons – more soluble. The effect is most marked in cold water and deep water, so these communities may well provide early warning of impacts likely to be seen later in other parts of the world's ocean. Riddle also suggests that increased understanding gained from this research of the association between fragile seabed communities and fish of commercial value might prove helpful in predicting the impacts of commercial fishing, particularly the destructive effects of bottom trawling or longlines laid along the seafloor.

Like his Larsen ice-shelf colleagues, Riddle postulates that scientists are only beginning to understand the complex biodiversity that lies beneath the surface of the Southern Ocean and its importance in local, regional and global ecosystems. He and his fellow explorers believe that their research is laying the foundation for scientists to understand how animal communities have adapted to the unique Antarctic environment, while also providing a framework for wider application of this knowledge as ongoing changes in the region accelerate.

POLAR OPPOSITES:
DIFFERENCES IN THE ICE OCEANS

Part of the sea-ice science team on the U.S. Coast Guard cutter Healy *is lifted to the ice in the Canada Basin during a Census expedition to this Arctic region in 2005.*

Antarctica and the Arctic are literally polar opposites, but geography is only one of many differences between these regions. Antarctica is a continent surrounded by ocean and virtually covered in ice, while the Arctic is an ocean surrounded by continents and by Greenland. The physical characteristics of these regions create differences in the sea ice found in them. The landmasses that nearly surround the Arctic Ocean serve as barriers to the movement of sea ice – hence, making it not as mobile as the sea ice surrounding Antarctica. Arctic sea ice does, however, shift and move within its ocean basin; the floes tend to bump into and pile up on one another, forming thick ice ridges.

These converging ice floes help to make Arctic ice thicker than the unconstricted sea ice found in the Southern Ocean off the coast of Antarctica.

The thickness of the Arctic ridge ice means that some of the ice stays frozen during the summer melts and continues to grow during the following autumn. It is estimated that of the 15 million square kilometers (5.8 million square miles) of sea ice that exist during winter, on average 7 million square kilometers (2.7 million square miles) remain at the end of the summer melt season. These figures are being watched closely as global temperatures warm.

Southern Ocean sea ice forms ridges much less often than sea ice in the Arctic. Also, not being

The Arctic Ocean, at the top of the planet, is nearly surrounded by land, which restricts movement of sea ice outside the ocean basin.

A female polar bear and her cub pay a visit to Census researchers aboard the U.S. Coast Guard cutter Healy. The curious bears came within 200 meters (660 feet) of the ship.

There are 17 species of penguins, six of which can be found in Antarctica. This gentoo penguin was photographed during one of 18 Census expeditions to Antarctica during the International Polar Year, March 2007–March 2009.

bounded by land to the north, the ice floats into warmer waters, where it eventually melts. Unlike in the north, almost all the sea ice that forms during the Antarctic winter melts during the summer. During the winter, up to 18 million square kilometers (6.9 million square miles) of ocean is covered by sea ice, but by the end of summer only about 3 million square kilometers (1.1 million square miles) remain.

Because the sea ice in the waters surrounding Antarctica is newly made each year, it is also less thick than Arctic sea ice. Antarctic ice is typically 1 to 2 meters (3–6 feet) deep, as opposed to Arctic sea ice, which most often reaches depths of 2 to 3 meters (6–9 feet), and in some areas up to 4 to 5 meters (12–15 feet).

Land inhabitants also vary between these two regions. For example, polar bears live only in the Arctic. No terrestrial mammals live in Antarctica — except for the occasional human researcher or tourist — while the land surrounding the Arctic Ocean hosts several: reindeer, wolves, muskoxen, hares, lemmings, foxes and humans. The Arctic is home to more than a hundred bird species, while less than half that number choose to live on the southernmost continent of the planet.

Regardless of their many differences, from ice composition to the species that inhabit them, the polar ice oceans share a common challenge: adapting quickly as temperatures rise and conditions change in these fragile ecosystems.

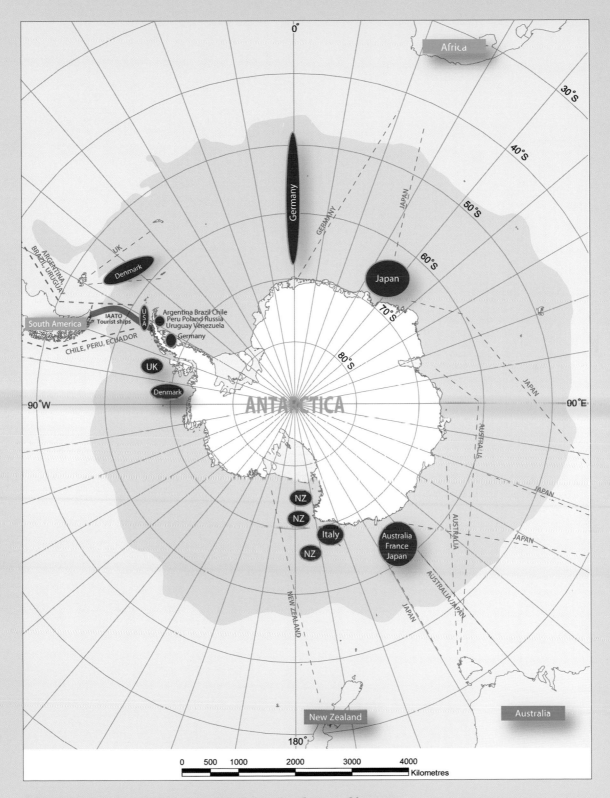

Eighteen research expeditions contributed to the Census of Antarctic Marine Life during the International Polar Year (2007–09), greatly increasing knowledge about this little-explored region.

CHAPTER SIX

Unexpected Diversity
at the Edges of the Sea

*Certainly what I intend to do with my life is record as much about
the planet as exists today as possible so that future generations have
a reference point of what the Earth looked like before it got
irrevocably altered by human activities.*

— GUSTAV PAULAY, CURATOR/PROFESSOR,
FLORIDA MUSEUM OF NATURAL HISTORY, AND PARTICIPANT IN THE
FRENCH FRIGATE SHOALS EXPEDITION, CENSUS OF CORAL REEFS

Interaction with the ocean has shaped the development of human societies and affected the history of the world. The ocean has long served humankind by helping to provide – up to now – a relatively stable climate, food, medicine, a platform for commerce and a place for recreation. It has also been, among other things, a major pathway for human migration and a vehicle for exciting discoveries. Despite this long history, the extent of humankind's knowledge about the ocean is not as great as one might expect – less than 5 percent of the bottom of the world ocean has been explored. Although the Census is cataloging methodically the diversity of life in the world ocean, discoveries of new species, habitats and communities sometimes happen completely by accident, just by being in the right place at the right time. This is true not just in the inaccessible reaches of the deep sea, but also in the nearshore environments where humankind has its most direct interaction with the sea.

New discoveries are constant reminders of how little is known about life in the ocean. Because their areas of study have been little explored, one would expect researchers working in the far-flung "special" environments

Opposite: A pre-Hispanic megamidden, or shell mound, of queen conch (Strombus gigas) *sits on La Pelona Island in the Los Roques Archipelago of Venezuela. At least 5.5 million conchs were exploited in this archipelago between 1200 and 1500 AD. Living in the shallow sandy areas that fringe sea-grass beds and coral reefs, queen conchs are easily harvested, as evidenced by the massive impact that even precolonial peoples had on their population. Heavy exploitation of the species in the area continued until the Venezuelan government finally banned the fishery in 1991.*

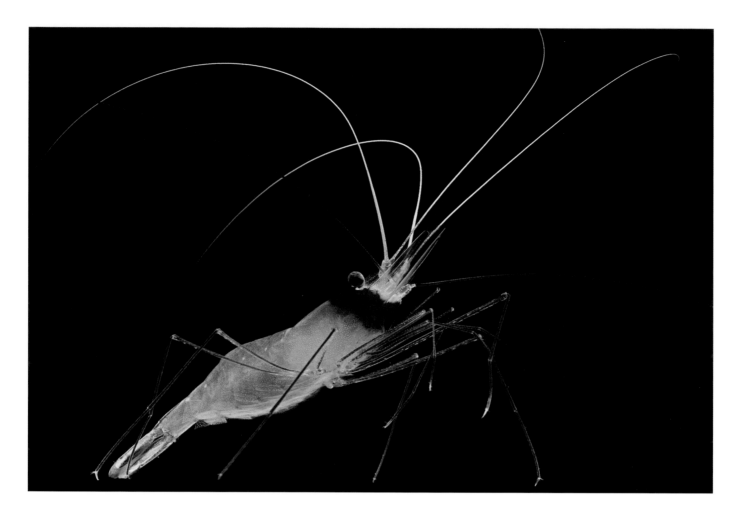

This deepwater shrimp, originally identified as Plesionika chacei, *may be a new discovery for nearshore Hawaii. Scientists are conducting further research to confirm its identity and determine whether there is previous evidence of its existence in the Northwest Hawaiian Islands.*

of the ocean – the abyssal plains, hydrothermal vents and deep ocean trenches – to encounter surprise discoveries. But researchers working in the most heavily affected nearshore areas are also finding unexpected communities, and they are still trying to identify a reference point or baseline for many nearshore ecosystems. Their discoveries illustrate the value of Census research and justify the sense of urgency with which marine and coastal scientists are working to fill in the unknowns. Without knowledge of what has lived and what currently lives in these nearshore environments, there is little hope of adequately conserving and protecting the quality of those ecosystems and thus sustaining the benefits they provide to humans every day. A pressing task is to identify the baseline from which future changes in the nearshore can be measured – the challenge being tackled by the Census of Marine Life.

The Census of Marine Life has three field projects that focus on the ocean realm termed the "Human Edges" – the nearshore environment. A handful of other projects in the Census also have programs within this realm, though their main focus may be broader or may primarily target other realms. The nearshore field projects base their research on the relatively thin band of ocean space that is closely associated with landmasses and therefore directly affected by the people who occupy the adjoining land. Both our dependence on these areas – notably the coral reefs, shorelines and coastal gulfs and seas – for ocean resources and the relative ease with which we can study them, compared to other ocean realms, have focused a large proportion of oceanic research on these nearshore environments.

With much still unknown – and potentially unknowable – about the global ocean and its inhabitants, including the species on which people depend for food or services and those prized for their aesthetic value, humankind's ability to conserve and preserve ocean life can be called into question. The Census is forging a foundation of knowledge in an attempt to provide a valid point of reference for future scientific and management efforts.

CORAL REEFS IN THE NORTHWEST HAWAIIAN ISLANDS

Ocean sample collection relies on innovative methods to collect organisms from the seabed, such as this airlift used to gently vacuum the ocean floor.

Census scientists investigating coral reefs are assessing the diversity and species composition of these threatened ecosystems worldwide, beginning with the Hawaiian Islands and working outward across the globe. In October 2006 the first expedition was launched to survey the coral reef environment of the Northwest Hawaiian Islands Marine National Monument, an untouched marine wilderness and the world's most remote island sanctuary. A group of Census researchers from a diverse mixture of universities, government agencies and museums spent three weeks investigating the marine environment surrounding French Frigate Shoals in this remote archipelago. The aim of the expedition was to explore the biodiversity of the region while simultaneously targeting understudied species groups in hopes of expanding knowledge of their ecology and taxonomy.

Census researchers take great pains to use techniques for sampling the diversity of reefs that do not destroy the reefs themselves. Using a mixed bag of collection methods in this location that included suctioning samples from the seabed,

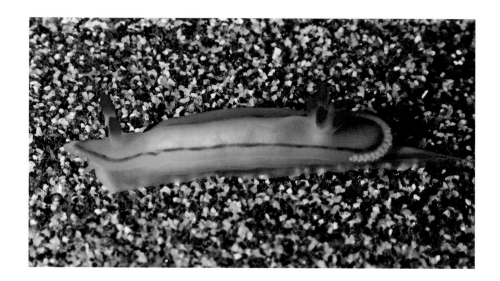

This opisthobranch, or sea slug (Thurunna kahuna), is showing its poisonous defensive mantle glands (the frilly crescent at one end), which use toxic secretions to discourage predators. This specimen was found during rubble extraction off Keehi Lagoon, Hawaii.

sifting through coral rubble with paintbrushes, and deploying traps, scientists collected approximately 4,000 samples, including more than 1,200 DNA samples to complement taxonomic analyses. After the samples were sorted, scientists estimated that more than a hundred new species of crustaceans, corals, sea squirts, worms, sea cucumbers and mollusks could be described from this work. Another result of the expedition was expansion of the known range of distribution for many species. For instance, at least 18 species of corals that were previously unknown in the area have now been observed and recorded in French Frigate Shoals.

In addition to identification of many new species and expanded ranges for existing species, the Northwest Hawaiian Islands expedition left one other lasting legacy. In 2006 the researchers deployed autonomous reef-monitoring structures, which were later collected in 2007. These structures, dubbed "dollhouses," were designed to provide a habitat for reef creatures that mimics the reef's structure. The aim was to provide a method for studying reef recolonization that would allow researchers to observe and interpret how reefs rebound from disturbances such as pollution events, storms and ship groundings. Such information can provide valuable guidance for future management and conservation of coral reef environments.

THE GULF OF MAINE: PAST AND PRESENT

The Gulf of Maine is one of the most heavily used ecosystems in the world. Throughout its history it has provided food, mineral resources and a transportation platform to those who live around its shores. Because it plays such an important role in the daily lives of so many

*Census researchers deploy an autonomous reef-monitoring structure (ARMS)
in the Northwest Hawaiian Islands. The structure, made of PVC, is designed
to mimic the nooks and crannies of a natural reef. An ARMS is used in studies
that test how marine life is recruited to freshly disturbed seabed space.*

people, the Gulf of Maine has also been the subject of much scientific study. Through the Census of Marine Life's work in the Gulf, understanding of its biodiversity continues to grow beyond expectations. Originally projected at approximately 2,000 species, the list of creatures that call the Gulf of Maine home now stands at more than 3,200 and continues to grow. Much is being learned through collaborations among a host of Census researchers.

The Gulf of Maine project is investigating the shoreline and nearshore areas of the Gulf from two perspectives: through a standardized shoreline biodiversity study and through a detailed historical investigation. Researchers in Maine are gaining a complete picture of the biodiversity of the area and how it has changed in recent history.

Studying the history of nearshore biodiversity is providing a historical context for the region's current biodiversity so that projections of future biodiversity can be made. Both the United States and Canada are involved in this effort; its aim is to apply ecosystem-based management – management that takes into account the cascading effects of alteration to or harvest from the ecosystem on the Gulf of Maine. A clear historical perspective such as that provided by Census research will likely prove to be an important tool in this endeavor, which includes resolution of any issues that might arise from transboundary disputes over resources.

Left: Scientists must also be accomplished divers to conduct underwater research in the murky nearshore zone of the Gulf of Maine.

Opposite: This pontoniine shrimp, collected in the Northwest Hawaiian Islands, is less than 2 millimeters long but is equipped with very large claws, or chelae. Scientists have not yet identified why a tiny shrimp would need such big claws, but they assume that it, like most shrimp, uses its chelae for food acquisition, mate attraction and defense.

JOINING FORCES

The Census nearshore sampling project is focused on investigating the biodiversity of shorelines and associated shallow-water habitats. This project is unique in that it is a truly global effort, with study sites in more than 45 countries and territories. It is also unique in that it is a grassroots-level endeavor in many countries, recruiting local researchers, students and public volunteers to assist in scientific data collection. By using local help and a standardized investigation protocol, the project is able to replicate its scientific methods over many research sites to gain the broadest perspective possible. This innovative collaborative approach to science has not only contributed to knowledge of the biodiversity of the world's nearshore areas but has also exposed many people of various backgrounds to marine science. In so doing it has changed lives through increased understanding and appreciation of marine life at and below the water's surface.

The August 2007 expedition to Zanzibar. U.S. and Kenyan students join forces.

In 2004 a group of high school students from Niceville, Florida, on the southeastern shore of the Gulf of Mexico, and a group of high school students from Wakayama, on the Pacific coast of Japan, participated in an exchange program to investigate nearshore biodiversity on two continents. First the students from Florida formed a Census investigative node and began sampling the Gulf of Mexico shore near their homes. Later that year they traveled to Japan, where they joined the Japanese students to take part in a nearshore research workshop at the Seto Marine Science Institute to learn sampling and analysis techniques.

In 2006 another high school group, the Kesennuma High School biology club, from Miyagi Prefecture, Japan, joined the Niceville and Wakayama students in their international nearshore sampling effort. These student groups have since continued their work, sampling and investigating the nearshore in their home areas and contributing their data to the Census effort.

In 2007 the Niceville students launched an even more ambitious endeavor: an expedition to Africa to teach marine science and to participate in sampling the first island site in the Indian Ocean — Zanzibar. Joining forces with Kizimkazi High School students from Kenya's Central Province, local university students, and African and American researchers, they sampled the shores of Tanzania's fabled island. The focus of the expedition was to initiate baseline sampling in the region and to encourage the local high school students to start their own Census nearshore sampling nodes, much like those established in Niceville, Wakayama and Kesennuma. This exchange program has provided research and enrichment opportunities for high school students around the world that will have lasting positive effects on both the students and the environment. By recruiting students and the public to get their hands dirty and participate in real scientific research, the Census is helping to create informed citizens who have more curiosity about and a better understanding of what lives near the water's edge.

A NEW HABITAT FOR ALASKA

When sampling for NaGISA in Prince William Sound, my co-worker dropped a sieve over the side of the boat on which we were sorting our samples. We did a dive to retrieve the sieve (60 feet [18 meters]) and found a totally new habitat for our state. We now know that there are rhodolith beds in Alaska.

— BRENDA KONAR, PROFESSOR, UNIVERSITY OF ALASKA FAIRBANKS, AND CENSUS OF MARINE LIFE RESEARCHER

Nereocystis, *a marine alga commonly referred to as bull kelp, is often found in the nearshore and shallow gulf areas of the Pacific coast of North America.*

The discovery of rhodolith beds in Alaskan waters, as described above by Brenda Konar, illustrates how Census work in nearshore environments is shaping the future of marine science and its applications. Rhodoliths are unattached chalky red algae that form dense beds on the seafloor. The Alaska discovery is significant because rhodolith beds around the world occupy an important niche, providing indispensable services within various ecosystems. These services include acting as a protected nursery for many marine species and a critical habitat for important commercial species, which can include clam and scallop varieties. Scientists are unsure how this newly discovered habitat relates to the surrounding ecosystem, or what role and importance it may have in the health and functioning of Alaska's nearshore environments. It is another example of how more questions are raised as more is learned about what lives below the surface of the ocean.

Rhodoliths are distributed by ocean currents, rolling like miniature tumbleweeds across the seabed. They are associated with a host of other species that use their structures for habitat, so their recent discovery in Alaska could signal the presence of other species yet to be documented in these waters. This new information is likely to spark debate over use of the seabed in Alaskan waters. Current legislation and regulations may not offer enough management or protection to conserve rhodolith beds, which, considering their recent discovery, may be relatively rare in Alaska. Without a clear understanding of how these beds fit into the overall ecological picture and what roles they play or services they provide in the local environment, management of the seabed and its uses could be a contentious issue among shellfish harvesters, conservationists and other stakeholders. The continued work of Census researchers may help to provide the information necessary for appropriate management of this habitat and the issues associated with it.

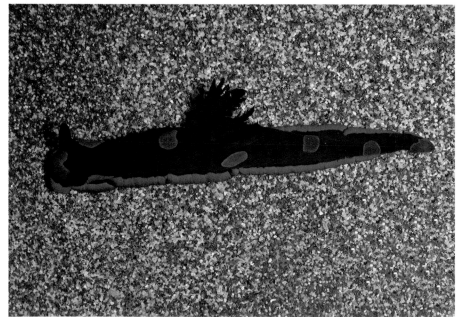

Above: A striking splash of color in the Arctic nearshore environment, this anemone is one of the species encountered by Census researchers as they surveyed the shallows close to the shoreline.

Right: Collected off of Keehi Lagoon, Hawaii, this sea slug (Tambja morose) *feeds on bryozoans (colonial coral-like animals) that it encounters in its coral rubble habitat.*

PROTECTING CORAL REEFS

An accelerated decay in biodiversity due to interacting human threats has been long suspected [by] ecologists and conservation biologists. Now we have an idea of the speed at which populations can decay when exposed to several threats.

— CAMILO MORA, POSTDOCTORAL FELLOW AT DALHOUSIE UNIVERSITY AND CENSUS OF MARINE LIFE RESEARCHER

A member of the Ovulidae (egg cowrie) family of sea snails (Primovula beckeri) feeds on the polyps of corals and sea fans. This ecological association is so specific that the mollusk has developed an ability to mimic the coloration and even the texture of the coral species on which it lives.

Coral reefs, the rainforests of the sea, have long been touted as a bastion of biodiversity within the world ocean. Indeed, in the relatively barren tropical seas, coral reefs are oases of productivity. They often support commercially important fisheries, can drive the economies of some nations through tourism, and are ecologically critical habitats in tropical oceanic systems. Despite their biological, economic and aesthetic importance, however, fewer than 20 percent of tropical coral reefs worldwide fall within marine protected areas (MPAs).

Census researchers studying the future of marine animal populations have concluded that of the 18.7 percent of coral reef habitats located within MPAs worldwide, fewer than 2 percent are afforded adequate protection to prevent further degradation. This inadequate protection is due largely to insufficient legislation and enforcement to prevent species harvest, poaching and other destructive uses. Even more disturbing is the fact that most inadequately protected coral reef habitats tend to be in regions that support the highest biodiversity, such as the Indo-Pacific and the Caribbean.

While MPAs exist in most regions of the world ocean that support coral reefs, the quality of these protected areas varies greatly. When they analyze MPAs for potential effectiveness in conserving coral reef biodiversity, researchers conclude that the global network of protected zones is troublingly insufficient. With coral reefs in decline worldwide, the need for appropriate conservation measures has never been more crucial.

While the major focus of MPAs worldwide is reduction of human pressure on the ecosystem, many of them remain "multiple use" protected areas that still allow harvesting and other potentially destructive activities. Considering that many scientists believe the amount of protected coral reef space worldwide is likely insufficient, even if the protected areas worked perfectly, the forecast for the reefs' future is not bright. Census findings point to the need for reevaluation of MPAs on a global scale – and the conservation strategies they support – to ensure that future conservation measures are more effective and that the benefits that coral reefs provide to the environment and to humans can be sustained.

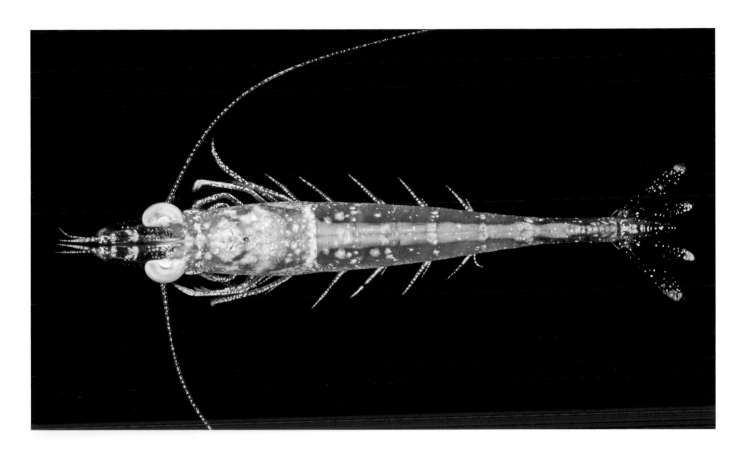

THE WAY FORWARD

The Census experience investigating nearshore areas reinforces how much still remains unknown about life in the world's ocean. The unexpected diversity that is being found in the familiar ecosystems of the nearshore, shallow gulfs and seas, and coral reefs is surprising even the most seasoned marine scientists. However, conservation and management success can reflect only what is known, proven and understood. There is a long road ahead to exploring, fully understanding and successfully stewarding the marine environment, the nearshore areas in particular. The Census is leading the way in providing the best possible starting point for the future of ocean and coastal management and conservation. By making discoveries around every corner, operating programs that involve schoolchildren and even big business in marine science, and integrating historical knowledge into its vision for the future, the Census of Marine Life is helping to close the gaps in our understanding of the nearshore and coral reef environments.

Above: This striking shrimp was collected from French Frigate Shoals in the Northwest Hawaiian Islands.

Opposite: This goggle-eyed worm belongs to the phylum Polychaeta, a group of segmented worms named for their many bristles (poly = many, chaete = bristle). Many species in this family undergo a spectacular transformation as they become sexually mature: both sexes develop huge eyes, while most of the body segments and bristles become paddle-shaped.

BIG BUSINESS INVESTS IN CENSUS RESEARCH

Crabs of the genus Trapezia, *such as this coral guard crab* (Trapezia cymodoce), *have a symbiotic relationship with their coral hosts. The crab defends the corals from predators and keeps them free from sediment in exchange for living space and food in the form of coral mucus.*

Census researchers are surveying the biodiversity and species composition of coral reefs worldwide. In conducting such surveys, the larger the study area, the better. The Australian energy and mining giant BHP Billiton has made it possible for Census investigators to cover more coral reef territory.

In a partnership between big business, science and conservation, the likes of which is becoming increasingly common globally, BHP Billiton, the Australian Institute of Marine Science and the Great Barrier Reef Foundation joined forces to ensure that Australian coral reefs were included in the Census of Marine Life's international efforts. Marine scientists and taxonomists are studying the two hallmark reef systems of Australia, the Great Barrier Reef off Australia's northeast coast and Ningaloo Reef on the west coast. A series of expeditions into the field for collection and observation, as well as follow-up lab work and analyses, is planned for a four-year period that began in 2006.

According to Ian Poiner, chief executive officer of the Australian Institute of Marine Science and chair of the Census's Scientific Steering Committee, "Scientists estimate that less than 10 percent of life on coral reefs has been identified. The Census of Coral Reefs is a major research effort to sample, analyze and document marine life on our reefs. This collaboration will also improve scientists' access to existing data and information about coral reef biodiversity, which will ultimately improve our understanding of coral reefs and how best to protect them."

The benefits will also flow in reverse. A research assistantship program offered by BHP Billiton will enable its employees to go into the field to help collect samples and survey reefs. This first-hand experience and close association with marine scientists will create a new force of coral reef stewards who better understand the complexity of these ecosystems and the challenges faced for their preservation. This partnership and the Census work that it supports will serve as a model for future scientific endeavors and will provide a foundation for the way forward.

A school of manini fish passes over a coral reef at Hanauma Bay in Honolulu. Many coral reefs are dying from water pollution (from sewage and agricultural runoff), dredging off the coast, sedimentation and careless collecting of coral specimens.

CHAPTER SEVEN

Unexplored Ecosystems: Vents, Seeps, Seamounts and Abyssal Plains

We know that seamounts support large pools of undiscovered species, but we can't yet predict what lives on the unstudied ones. The tragedy is we might never know how many species go extinct before they are even identified.
— FREDERICK GRASSLE, CHAIR OF THE CENSUS OF MARINE LIFE
SCIENTIFIC STEERING COMMITTEE AND THE FIRST BIOLOGIST
TO INVESTIGATE HYDROTHERMAL VENTS

Imagine a continent, with its vast plains, deeply gorged valleys and craggy mountaintops. Now imagine all that covered by water thousands of meters deep. The bottom of the ocean has many features analogous to continental landscapes, but it also has deep-sea geologic features that present unique challenges for the life that exists there. It has long been suspected that the abyssal plains (deep, vast expanses of relatively flat seabed), seamounts (underwater mountains), hydrothermal vents (fissures in the crust that emit superheated fluid) and continental margins of the world's ocean are home to many undiscovered species. Most of these regions have remained largely unexplored because of the difficulties and expense of reaching them, but now technological innovations have made it possible for ocean scientists to look into these relatively inaccessible habitats. Through daring expeditions and hard work, Census of Marine Life scientists have made major contributions to the understanding of life in these remote areas of the ocean.

Opposite: Light from a remotely operated vehicle allows the exploration of unique formations in the deep sea.

153

HYDROTHERMAL VENTS

Deep-sea hydrothermal vents and their associated fauna were first discovered in 1977 along the Galapagos Rift in the eastern Pacific. It is now known that these extraordinary seafloor hot springs are found along the 48,000 kilometers (30,000 miles) of mid-ocean ridges that form Earth's largest continuous volcanic system.

In vent systems, hydrothermal fluid – originating from seawater seeping through Earth's crust – re-emerges from the seafloor at temperatures of more than 350°C (662°F). In 2006, Census scientists discovered the hottest hydrothermal vent ever recorded; its temperature was 407°C (765°F), hot enough to melt lead. The fluid that is propelled from the vents contains dissolved metals and sulfur, which precipitate out when the superheated liquid meets the surrounding cold seawater, giving it the appearance of dense black smoke. The precipitated material forms vent chimneys

Smoky white vent fluid rises out of small sulfur chimneys at the Northwest Eifuku volcano in the Mariana Arc of the Pacific Ocean. This area was named the Champagne Vent because bubbles of liquid carbon dioxide were rising out of the seafloor.

that can reach as high as 20 meters (66 feet). Colonies of unusual marine life such as clams, tube worms and exotic microorganisms cluster around these vents, feeding on the chemical soup spewing from the ocean floor.

Census explorers of vents have discovered many new species from a number of different animal groups. Many are exclusive to these ecosystems and would be unable to exist elsewhere. The composition of the animal communities found at different vent sites has also been found to be distinct. For example, the vents in the East Pacific are dominated by giant tube worms (*Riftia*), large white clams (*Calyptogena magnifica*) and mussels (*Bathymodiolus*). In the Atlantic, however, the vents are dominated by dense aggregations of shrimp and mussel beds. The recently explored Indian Ocean vents had some surprises to offer: while most of the fauna is related to animals of the Pacific, the dominant species is the common shrimp *Rimicaris* – from the Atlantic.

These Bathymodiolus *mussels and an unidentified fish live near hydrothermal vents on the Mid-Atlantic Ridge.*

HYDROTHERMAL PROCESSES

GAKKEL RIDGE

Hydrothermal vents have been discovered on the Gakkel Ridge, at the bottom of the Arctic Ocean.

To measure growth, scientists stained these tube worms (Lamellibranchia luymesi), found in the Gulf of Mexico at a depth of 540 meters (1,775 feet), with blue dye. The white area indicates new growth over one year. From these studies scientists discovered that these tube worms can live for more than 250 years.

The Gakkel Ridge in the Arctic Ocean is the slowest-spreading mid-ocean ridge on Earth. During a ground-breaking research expedition in 2001, new hydrothermal vents were discovered there – nine vent sites close together, with at least one site every 100 kilometers (160 miles).

In 2007 an international team of Census scientists returned to Gakkel Ridge to further study its vent communities. The scientists were armed with new equipment that included a towed real-time imaging and benthic sampling system called *Camper* and two new autonomous underwater vehicles (AUVs), *Puma* and *Jaguar*, which were specifically designed for locating and sampling vent systems and fauna in the Arctic Ocean. *Puma* ("plume mapper") used temperature and chemical sensors and a novel laser-guided optical sensor to search for vents. Once a venting area was located, the AUV *Jaguar* was used to conduct a high-resolution bathymetric survey of a small area and to collect magnetic data and images of the biological communities in the vent areas. *Camper*'s high-definition imaging, sampling and sensing capabilities added to the information by providing high-resolution seafloor images and collecting bottom samples, using either a clamshell "grab" sampler or a suction "slurp" sampler.

The major accomplishments of the 2007 Gakkel Ridge expedition included discoveries of the Asgard volcanic chain, extensive chemosynthetic microbial mats covering the volcanoes, and basaltic glass fragments over large portions of the seafloor, providing evidence of explosive volcanism. Techno-logical advancements emerging from this research included detailed mapping of water-column plumes, development and demonstration of *Camper*'s wire-line system for high-resolution imaging and sampling of the deep seafloor under ice, and development of the AUVs *Puma* and *Jaguar*, which could operate from deep within the ice pack.

Ed Baker was a participating oceanographer on the 2007 Gakkel Ridge expedition and is a supervisory oceanographer at the National Oceanic and Atmospheric Administration (NOAA) Pacific Marine Environmental Laboratory in Seattle, Washington. He has been studying hydrothermal vents for almost 20 years, and states that the Gakkel Ridge discovery was among the most remarkable and unexpected of his career. "This discovery is significant because it is so unexpected," Baker says. "The tectonic plates on either side of the Gakkel Ridge spread apart, or open, very slowly. In fact, it's the planet's slowest-spreading ridge, moving at about half an inch a year or less. We expected to find no more than four or five vent sites because this sluggish spreading rate creates far less volcanic activity than on most mid-ocean ridges."

After photographs from a towed camera showed shimmering water and abundant biological activity, the scientists named the area Aurora. The unusual marine life-forms found in Atlantic Ocean vent sites are "markedly different," says Baker, from those found in Pacific Ocean vent sites. Baker and his colleagues suggest that since the Gakkel Ridge is not connected to other parts of the mid-ocean ridge system south of Iceland, it is likely that new species of marine vent life await discovery. "We are eager to return and see what is living down there," he says.

One of the striking characteristics of the fauna living at vents is that the organisms live independently of the sun and its life-giving energy. Instead of photosynthetic plants, microscopic bacteria act as the fuel at the base of the food web in this extreme habitat. The bacteria, which produce organic compounds from carbon dioxide in the water, support dense populations of exotic organisms.

Because of the extreme conditions of the vent habitat, Census researchers hypothesize that certain species may have specific physiological adaptations – with interesting implications for the biochemical and medical industries. It has also been suggested that hydrothermal vents could be the type of habitat where life originated. In collaboration with NASA, these hypotheses are being used by Census researchers to develop programs to search for life in outer space.

In addition to hydrothermal vents, Census scientists have been actively researching another unique habitat known as cold-water seeps (see below). The ecosystems of hydrothermal vents and cold seeps are similar in that they do not rely on photosynthesis for food and energy production. They are called chemosynthetic ecosystems, because the life found there depends on chemical processes rather than photosynthesis.

At a thermal vent 3 kilometers (5 miles) below the surface in the equatorial Atlantic, Census researchers found shrimp and other life-forms on the periphery of fluids billowing from Earth's crust. The fluid temperature was an unprecedented marine recording of 407° C (765° F) – hot enough to easily melt lead.

CONTINENTAL MARGINS AND COLD-WATER SEEPS

A continental margin is the slope that extends from the edge of the continental shelf – about 200 meters (660 feet) from shore – to the abyssal plains, which are 4,000 to 6,000 meters (2.5–3.75 miles) below the water. The Census is establishing biodiversity baselines in continental margin areas worldwide that are still untouched by commercial exploitation, as well as collecting evidence of change in areas with commercial activity. Oil exploration is a major international interest in continental margin areas, and with increasing pressure for oil resources, it is important to understand these relatively unexplored ecosystems.

Census scientists have discovered that, during the past few decades, deep continental-margin habitats have changed more than any other large area of Earth. Once envisioned as monotonous landscapes, continental margins are now acknowledged to have a high degree of complexity and diversity. Fundamental patterns of species distribution that were first observed and explained in the context of monotonous slopes must now be reevaluated in light of the new discoveries.

Above: The animals most commonly found at cold seeps are clams, mussels and tube worms. The bushlike clumps of tube worms provide a home for other organisms such as crabs and a variety of sponges, bryozoans and bristle worms.

Below: The tube worm (Lamellibrachia luymesi) *and its bacterial symbionts live on sulfides produced by anaerobic oxidation of oil and gas. Many oil and gas seeps in the Gulf of Mexico feature dozens of dense, bushlike aggregations of tube worms; this "bush" shows the red gill-plumes of several worms.*

LIFE IN A COLD-SEEP ENVIRONMENT

This species of hermit crab from a cold-seep site off New Zealand is as yet unnamed. Note the furry-looking seep-associated bacterial filaments on its claws.

These sea whips, or bamboo whip coral, live near brine seeps in the Gulf of Mexico's Bright Bank.

Brine seeps create a ghostly effect in the Gulf of Mexico.

Several animals have developed highly specialized relationships with cold-seep bacteria. One of these, a clam, obtains its food from the bacteria. How does this happen? The clams and the bacteria live together and help each other in an exchange process called symbiosis. In the cold-seep community, bacteria make their home inside the clam's gills. Clams have a muscular foot that helps them attach to the seafloor, and this foot takes in hydrogen sulfide from the water of the cold-seep area. The hydrogen sulfide, produced by methane-using microbes found in the seep, is carried by the clam's blood to its gills, where the bacteria live. The bacteria use chemical energy found in the hydrogen sulfide to combine carbon dioxide and water, creating sugars and other compounds needed for growth. The bacteria and the sugars released then become food for the clam.

Likely because of the cooler temperatures and the stability of their environment, many cold-seep organisms are much longer-lived than those inhabiting hydrothermal vents. Recent research has revealed that the seep tube worm (*Lamellibrachia luymesi*) may be the longest-living noncolonial invertebrate known, with a lifespan between 170 and 250 years.

Cold seeps develop a unique topography over time. They can become littered with carbonate rocks ranging from small pebbles to huge blocks, which are thought to be formed by byproducts of microbial metabolism precipitated from seep waters.

"Ice worms" (Hesiocaeca methani-cola) *live in methane hydrates in the northern Gulf of Mexico.*

Quite recently submersibles have provided access to biodiversity hot spots on continental margins that would be considered some of the best mountain scenery on Earth if they were not covered by water. Cold seeps – areas where hydrocarbons such as methane and oil ooze out of sediments in the ocean floor – are among the hot spots. These areas are home to many species that have not been observed anywhere else on Earth. Cold seeps have now been found along both active and passive continental margins at depths ranging between 400 and 8,000 meters (1,300–26,400 feet), including the Monterey Canyon just off Monterey Bay, California; in the Sea of Japan; off the Pacific coast of Costa Rica; in the Atlantic Ocean off Africa; in waters off the coast of Alaska; and under an ice shelf in Antarctica. The deepest seep community known is in the Japan Trench, at a depth of 7,326 meters (4.6 miles).

Bacteria harvest the chemical energy from hydrogen sulfide or methane found in the seep fluids to produce sugars, proteins and other building blocks of organic tissues. Various marine animals feed upon these bacteria, and Census scientists have discovered many new species living in these chemosynthetic ecosystems. One new kind of animal, *Hesiocaeca methanicola,* recently discovered on cold seeps in the Gulf of Mexico at about 500 meters (1,650 feet), was named the "ice worm" by Census researchers. It lives in extensive burrows that it excavates in seafloor deposits of gas hydrates – natural formations of ice crystals with gas inside. Scientists have analyzed the stomach contents of these worms and have found sediments and large bacterial cells that could have been obtained by grazing hydrate surfaces. However, there is still much more to learn about the nutrition and life history of this animal.

Census scientists have developed a long-term international field program for discovering and exploring new vent and seep sites and studying their fauna. The selection of key locations has been based on a number of specific scientific questions about distribution, isolation, evolution and dispersal of deepwater species from chemosynthetically driven systems. Two large areas of combined systems have been targeted for exploration: (1) the Equatorial Belt, from the Costa Rica cold seeps to the African continental margin, including the Cayman Trough, the Gulf of Mexico cold seeps, the Barbados Prism, the Mid-Atlantic Ridge north and south of the Romanche Fracture Zone, and the northern Brazilian continental margin, and (2) the Southeast Pacific, including the Chile

Rise, which is the cold-seeps and oxygen-minimum zone of the southern Chilean continental margin – an important migration area for whales. Also on the Census's list for research are a number of locations that already have important national and/or international support. These areas, grouped into three regions from pole to pole, comprise the Gakkel Ridge in the Arctic, the East Scotia Ridge in the Antarctic, and the Central and Southwest Indian ridges in the Indian Ocean.

SEAMOUNTS

Seamounts are literally undersea mountains, which are usually defined as fully submerged mountains or hills rising 1,000 meters (3,300 feet) or more from the ocean floor. They are found throughout the world's oceans, many in international waters, where they are governed by a complex array of multinational treaties. Nearly half of them are in the Pacific Ocean, while the rest are found mostly in the Atlantic and Indian Oceans; overall, more are found in the Southern Hemisphere. Seamounts are often isolated, which contributes to the uniqueness and diversity of their ecosystems. Usually cone-shaped, they are often volcanic in origin – forming near mid-ocean spreading ridges, over upwelling plumes caused by rising seawater, and in island-arc convergent settings. Oceanic islands are seamounts that have breached sea level.

Census scientists estimate that fewer than 400 seamounts have been sampled, and of these, fewer than 100 have been sampled in any detail. A key aim of the Census is to increase the number of seamounts that have been explored and to ensure that they are sampled in sufficient detail to enable meaningful conclusions to be drawn about the life found there.

Seamounts can be hot spots of marine life in the vast expanses of the oceans, and the species found on them differ from those found on the surrounding deep ocean floor. Some seamounts have been shown to support high levels of biodiversity and unique biological communities. In some instances, high levels of endemic species (that is, those only found at that locality) have been found. Seamounts may act as regional centers of speciation (places where new species emerge), as stepping stones for dispersal across the oceans, and

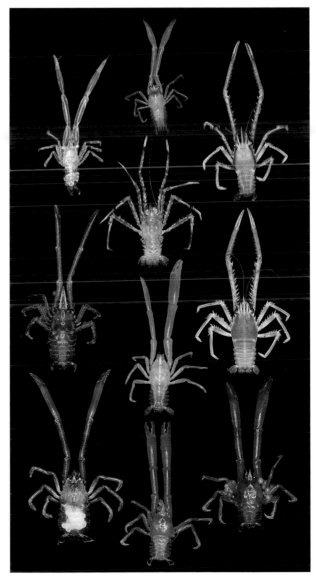

The diversity of life on a seamount is illustrated by the variety of these squat lobsters from two sister families of crustaceans, Galatheidea *and* Chirostylidae. *These specimens were caught off the seamount chains north of New Zealand in 2003.*

as places of refuge for species with a shrinking range. Census scientists have found that seamounts are home to an astonishing diversity of species, with 40 percent endemic to each mountain. Thousands of new species have been discovered in recent years – 600 on just five seamounts!

To date the most dominant organisms found on the hard surfaces of seamounts are suspension feeders – animals that rely on water to deliver oxygen and food – such as coral, sponges and sea fans. Soft sediments also accumulate on seamounts; the dominant organisms occurring there are polychaetes, a type of marine worm. Other animals that live in the sediments include oligochaetes (another type of worm) and gastropod mollusks (shelled animals).

Above: Stunning yellow Enallopsam- mia *stony coral live on Manning Seamount along with pink* Can- didella *teeming with brittle stars.*

Seamount species are sustained by food carried by passing currents. Nutrient-rich water is deflected upward by the mountain's slopes, picking up speed as it flows over the summit. Close to the summit, thriving communities of suspension feeders filter organic matter from the passing water. Fish feed on prawns, squid and small fish that drift by, while sea spiders and lobsters find refuge in the coral and rock outcroppings, and bottom-dwelling animals benefit from nutrient fallout from above. Whales and tuna visit these undersea mountains on their migratory routes. Further down the seamount slopes, coral communities become more sparse – a phenomenon akin to the tree line on terrestrial mountains, but in reverse.

Census scientists aboard the RV *Tanagaroa* on a month-long voyage to survey the Macquarie Ridge discovered millions of brittle stars catching passing food in a rattling current of 4 kilometers per hour (6.4 mph), and dubbed the place Brittlestar City. Its cramped inhabitants, tens of millions living arm-tip to arm-tip, owe their success to the seamount's shape and to the swirling circumpolar current flowing over and around it. The current allows the brittle stars to capture passing food simply by raising their arms, while it sweeps away fish and other hovering would-be predators.

Seamounts also provide habitats and spawning grounds for larger animals, including numerous fish. Some species, such as black oreo and black cardinalfish, occur more often on seamounts than above the adjacent slopes and seafloor. Nearly 80 species of fish and shellfish are commercially harvested from seamounts; they include rock lobster, mackerel, deep-sea red king crab, red snapper, several tuna species, orange roughy and perch.

A beautiful, as yet unnamed, white sponge with purple crinoids was discovered on Retriever Seamount.

This species of coral is another example of the colorful marine life found on Retriever Seamount.

Spiraling Iridigorgia *corals, sharing their space with brightly colored commensal shrimp, were discovered living on the New England seamount chain.*

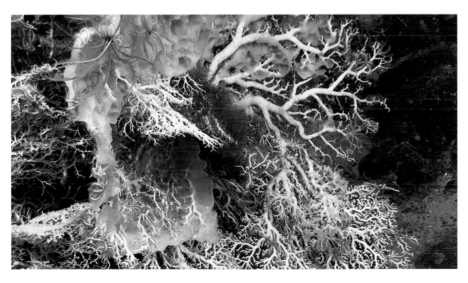

An example of the abundance and diversity of colorful species found on seamounts in the Pacific Ocean is a bouquet of Corallium *coral with deep purple* Trachythela octocoral, *plus brittle stars, crinoids and sponges.*

Seamount communities are under threat from overexploitation and physical damage from deep-sea trawling. Census scientist Malcolm Clark, of New Zealand's National Institute of Water and Atmospheric Research, says, "Seamounts can be productive habitats and benthic fauna and fish species can be abundant in these areas. Deepwater fisheries for species like alfonsino, boarfish, cardinalfish and orange roughy, for example, often take place on and near seamounts, and as a result, the localized impact of trawling can be substantial. Very little is known of what actually lives on seamounts – only 300 to 400 have been sampled out of a possible global total close to 100,000. We need to learn more about the animals found on seamounts, how the communities are structured, how they function and how fisheries can affect them – so that seamount resources can be managed in a sustainable way and biodiversity is maintained."

Although many coastal seamounts have been fished for decades, researchers are warning that faltering fish stocks mean fishing fleets are heading into deeper waters in search of new catches. Boats are increasingly targeting pristine oceanic seamounts. Using sophisticated sonar equipment, trawlers are often the first to discover them, and bottom-trawling nets can do immense damage within a year or two. Studies in the Tasman Sea show that coral and crinoids (a group of suspension-feeding echinoderms) cover 90 percent of pristine seamounts. Once they have been fished by trawlers, that figure drops to 5 percent. After the area is abandoned by the trawlers, recovery of the ecosystem is painfully slow. Some seamounts in the North Pacific have still not bounced back some 50 years after boats first trawled them, says John Dower, a fisheries oceanographer at British Columbia's University of Victoria.

Census scientist Karen Stocks, from the University of California at San Diego, states, "These [seamounts] are not just threatened habitats; they support new and unique communities that can give us insight into the processes that create and maintain biodiversity in the oceans. In addition to their suspected role as stepping stones for the dispersal of species throughout the world's oceans, seamounts are also almost certainly centers of speciation – hot spots for the evolution of new species. In the last couple of years, major advances have really changed our perspective of how unique these habitats are, and how exciting and informative future research on them will be. Seamounts have also turned up discoveries of both very old life-forms – hundreds of years old for some corals and crinoids – and living fossils."

Left: A great diversity of organisms was found on Balanus Seamount off New England, as represented by a strange spoon worm, an elegant sea pen, a stalked crinoid and two xenophyophores (large, one-celled organisms) with brittle stars.

Below: Large primnoid corals, loaded with brittle stars, were observed on Dickins Seamount in the Gulf of Alaska.

This member of the phylum Mollusca (order Nudibranchia) was observed swimming on the flank of Davidson Seamount at a depth of 1,498 meters (4,943 feet) off California.

In the Tasman Sea, studies show that unfished seamounts have twice the biomass of fished ones, and only 10 percent bare rock compared to 95 percent on fished sites. The bare rock is evidence of a highly disturbed ecosystem. Fishing has been most intense on seamounts closest to the sea surface – within 650 to 1,000 meters (2,100–3,300 feet). There is growing concern among scientists and conservation groups that these habitats, potentially some of the most unique and prolific on the planet, may be irreparably damaged even before they can be discovered and studied.

ABYSSAL PLAINS

Abyssal plains are flat or very gently sloping areas of the deep ocean basin floor. They are among Earth's flattest and smoothest regions – and the least explored. Abyssal plains cover the vast majority of the ocean floor. Generally lying between the foot of a continental rise and a mid-oceanic ridge, they result when an uneven surface of oceanic crust (made of basalt) is covered by fine-grained sediment, mainly clay and silt. Much of this sediment is deposited by currents that have traveled down the continental margins, along submarine canyons (steep-sided valleys on the seafloor) and into deeper water. The remainder of the sediment is chiefly dust (clay particles) blown out to sea from land and the remains of small marine plants and animals (plankton), which sink from the upper layer of the ocean. The sediment deposition rate in remote ocean areas is relatively slow, estimated at 2 to 3 centimeters (about 1 inch) per thousand years. In some areas of these vast plains, manganese nodules are common, containing significant concentrations of metals that include iron, nickel, cobalt and copper. These nodules may prove to be a significant resource for future mining. Sediment-covered abyssal plains are less common in the Pacific Ocean than in other major ocean basins, because sediments carried by currents are trapped in submarine trenches that border the Pacific.

Because these wide expanses of ocean bottom are very difficult to reach, questions about how many species are there and how they are distributed have long remained unanswered. The Census of Marine Life has been the first large-scale effort to find some of these answers. Census scientists have collected hundreds of new species, and nearly 200 have already been described and named. Identification and description of deep-sea organisms is very important because, in any sample taken at any spot in the deep sea, at least half of all animals have never been seen before. At this time, estimates of the number of species living in the deep sea vary greatly from 500,000 to 10 million; such a wide range indicates the poor state of our knowledge. Before the Census began, scientists had to rely on a few football fields' worth of sampled seafloor to make assumptions about half of Earth's surface!

Research in these remote areas is fraught with difficulties. Not only does water depth need to be negotiated when sampling abyssal plains,

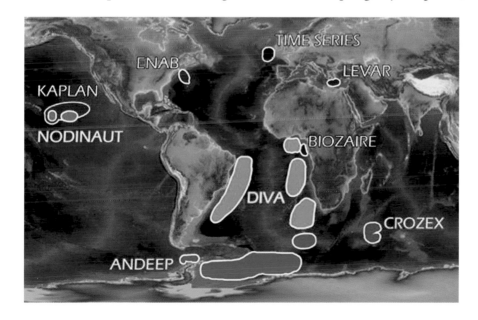

Census sampling sites are located throughout the world ocean's abyssal plains.

but the distance from land also makes access difficult and expensive. Often it takes several days for a ship to get to a deep-sea research area. The ships have to be large enough to withstand bad weather on the high seas and to accommodate enough crew and scientists for the challenging work on deck, making ship time very expensive. For example, maintenance of a ship such as the German icebreaker *Polarstern* costs about 60,000 euros (more than US$94,000) per day. Time becomes the most precious commodity on board a research vessel, and not every attempt to take a sample is successful. In the deep sea, eight or ten hours may pass before a scientist knows whether the sampling has succeeded or failed.

Census expeditions to deep abyssal plains have covered a wide variety of regions. In the Southern Ocean, research has focused on the evolutionary processes and oceanographic changes that have resulted in present biodiversity and distributional patterns, as well as on the importance of the Antarctic region as a possible source for deep-sea benthic species in other oceans.

The Congo canyon/channel system extends westward off the Congo-Angola margin of Africa for 760 kilometers (1,200 miles), sloping down to the abyssal plain at a depth of 4,900 meters (16,000 feet). A study of the canyon has looked at how contrasting environmental conditions – driven mainly by the activities of one of the world's largest and most active submarine canyons – might affect the biodiversity of the benthic communities there. Several types of equipment were used to collect information on the environmental parameters and benthic communities, including the French ROV *Victor*, which allows work on the bottom at a very precise scale. A Census study of the benthic communities at two sites 4,000 meters (13,200 feet) deep – one next to the Congo channel and influenced by its activities and the other 150 kilometers (240 miles) to the southeast, away from these influences – highlighted differences in community structures of the marine animals living there.

Another very ambitious Census study investigated the abyssal plains on both sides of the Atlantic Ocean in the Southern Hemisphere. In the course of these investigations, the scientists caught a glimpse of a hitherto unknown world. Many new and strange species of crustaceans and bristle worms were discovered; these new species now await descriptions and official names. Genetic analyses will help us understand the relationships between new and already known species.

Census scientists were delighted to find two quite unusual gigantic seed shrimps of the genus *Gigantocypris*. Seed shrimps (class Ostracoda)

An exotic species of the genus Bathypterois *was photographed in the Congo channel off Africa.*

are a common group of very small life-forms that can live in any aquatic environment, from the mud at the bottom of a small pond to the continental shelf. Seed shrimp live inside small, flattened shells made of chitin, like crayfish. Their shell is actually a combination of two shells attached like a clam shell. Seed shrimp are intense scavengers – the most common scavenger (based on quantities collected) in the sea.

The seed shrimps that are common in many parts of the ocean grow no larger than 2 to 3 millimeters, but the two individuals recently discovered reached nearly 3 centimeters; because of their orange-red color, they looked rather like cherry tomatoes. Other new species found were also much larger than normal for the same genus. How species richness in the deep sea develops and why gigantism is such a relatively frequent phenomenon in the deep sea, we can only guess. The knowledge gained by scientists during Census expeditions will help a great deal in answering these questions.

The abyssal plains are not just rich in marine life; they also hold potential for mining of metal resources. In the Central Pacific there is a vast area rich with manganese nodules. The Census is studying the

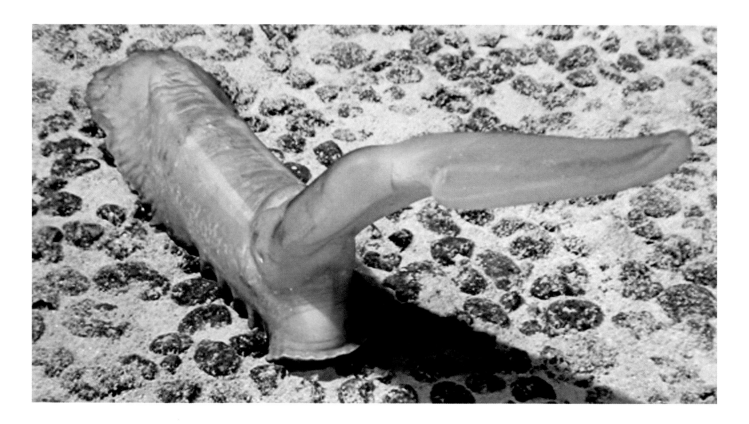

Above: A species of Psychropotes *moves along a bed of manganese nodules.*

Opposite: Nardoa rosea *sea star, as seen from the underside; Heron Island, Great Barrier Reef, Australia.*

region to determine what is living there and how habitat variability might influence its biodiversity.

An evaluation of the benthic community's recovery after dredging of nodules was conducted by sampling a dredge site that was about 25 years old and comparing it to untouched areas. This study revealed that the community structure of abyssal fauna in the nodule region differs from those in other areas, not only because of the availability and quality of food but also because of the physical and chemical properties of the habitat. It was documented for the first time that nodule fields constitute a distinct habitat for faunal communities and that faunal components differ in abundance depending on the presence of nodules. Interference with the habitat and mining of the nodules could affect the marine life there forever.

Census research in the deep, remote regions of the global ocean – the seamounts, abyssal plains, vents and margins – is revealing an incredible diversity and abundance of marine life. A wealth of new species is being discovered with every expedition. A better understanding of what is living where will aid in management of marine resources worldwide. Hopefully, that understanding will also be used to prevent further decimation of such critical ocean habitats.

CHAPTER EIGHT

Unraveling the Mystery of New Life-Forms

New species aren't really new, they are just new to us. These creatures have been out there for millions of years and we are just now fortunate enough to find them and have the technology to examine them.

— STEVEN HADDOCK, MONTEREY BAY AQUARIUM RESEARCH INSTITUTE, MEMBER OF THE CENSUS OF MARINE ZOOPLANKTON STEERING COMMITTEE

THE NAME GAME

Angelika Brandt has compared her work in the deep Southern Ocean to visiting a new planet, collecting the creatures that live there, and then trying to figure out how they all fit into their world. While trying to decipher the patterns and processes of the Southern Ocean, Brandt and her colleagues discovered more than 700 new species during three expeditions from 2002 to 2005. These discoveries gave the next round of researchers an opportunity to dig deeper to find answers to questions such as who eats whom, what each animal needs to survive, how and where they live, and how they interact with those around them.

"It was a total surprise for us to find so many new species, for we didn't expect to see such enormous diversity. Previous papers had documented that species numbers decline at lower latitudes, so while we thought we might find something exciting and new, we were totally overwhelmed by the fact that more than 95 percent of what we found were new to science," says Brandt, who is based at the Zoological Institute and Zoological

Opposite: This potential new species of amphipod crustacean was collected near Elephant Island during the 2006–07 Weddell Sea expedition. Scientists suspect that it is a new species and new genus.

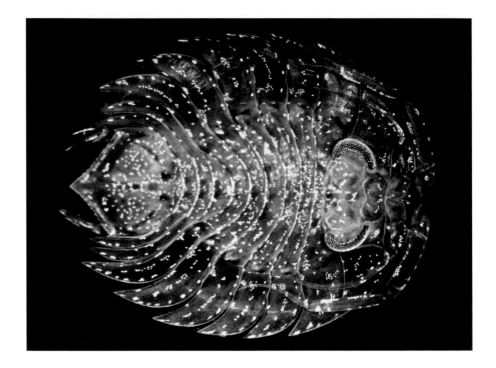

Found on the Weddell Sea slope, this serolid isopod has a flattened body that helps it dig quickly into the sediment. Aquarium observations have shown that only a small part of the animal's back sticks out of the sediment, allowing it to breathe.

Museum, University of Hamburg, and is a researcher with the Census deep-sea project that studies abyssal plains.

The discovery of that whopping number of new species in the Southern Ocean is really just the beginning. Census researchers are discovering new species at a much faster rate than they can be described. Since fieldwork began in 2003, Census of Marine Life scientists have discovered more than 5,300 species that are potentially new to science, ranging in size from 1-millimeter-long zooplankton to a giant Madagascar lobster weighing 4 kilograms (9 pounds). Between 2003 and 2008, however, only 110 of these creatures underwent the rigorous scientific review process required for official designation as a new species. This process can – and often does – take years.

The challenge for Census scientists and their colleagues worldwide is that before they can begin to seek answers to the many engaging questions posed by the discovery of a new species, they have to embark on (some might say endure) the process of ensuring that what they have found is indeed a new species. Some scientists compare it to the search for the perfect mate: researching everything that has ever been written about what constitutes a perfect mate, then interviewing your potential mate's friends and family to ensure that the prospect is exactly who he or she appears to be. In the case of a new species, after ensuring that it is not an impostor, next comes the process of deciding on a name and getting

Some of the deepwater antarcturid isopods found (such as this one, from the Weddell Sea) have eyes, suggesting that they evolved from species that lived on the shallower continental shelf, where light penetrates to the bottom. This is a juvenile of the genus Cylindrarcturus. *Scientists are not entirely certain that this is a new species because some of its characteristics were not completely developed.*

a publication interested enough in your newfound life-form to publish an article about it. After all this, a specimen of your new species must be deposited in a museum collection so that others can compare their discoveries to yours for years to come. Much like matchmaking in the modern world, declaration of a new species is anything but a simple process.

To identify a species never before cataloged requires significant effort, precise and often tedious attention to detail, and an inordinate amount of patience. The multi-step process begins the moment a specimen that is not readily identifiable is collected. After the initial euphoria of a potential new discovery, the long, thoughtful, laborious work

Researchers aboard the German research vessel Polarstern *sort through flora and fauna from a bottom sample as the first step of the species identification process. The sample was collected during an expedition to the Weddell Sea in 2006–07.*

Discovery of a new species of giant spiny lobster (Panulirus barbarae) *off Madagascar raised questions about the origins of the stock, its replenishment and how it will react to the inevitable fishing pressure that will follow its discovery.*

begins. Whenever possible the specimen is photographed, illustrated with line drawings and preserved so that it can be compared to similar specimens. In some instances, genetic material is also extracted and DNA sequencing is performed. The ultimate goal of the DNA research is to study different genes for analysis of phylogenetic relationships – the evolutionary history of a particular group of organisms – or to create a unique DNA barcode for the species so that it can be readily identified in the future.

The next step in species identification usually consists of consultation with colleagues to see if someone has come across an organism like the one in question. This is followed by a search of the taxonomic literature to ensure that the unidentified specimen has not already been found and described. This literature search must be exhaustive – and it can certainly be exhausting for the eager researcher. If other, similar specimens are available, time, money and effort can also be invested in visiting various museum collections to compare the new specimen(s) with others of the same type that may have been preserved for years. (Some archived specimens are more than 200 years old, dating back to when Linnaeus's species classification system was instituted in 1758.)

Once the researcher is confident that the specimen actually is a new species, the next step is to write a detailed description of the species and determine an appropriate name for it. Species' scientific names are always in Latin and consist of two words. The first word is the genus name and the second identifies the species within the genus. It is with the second word that creativity can enter the naming process. These

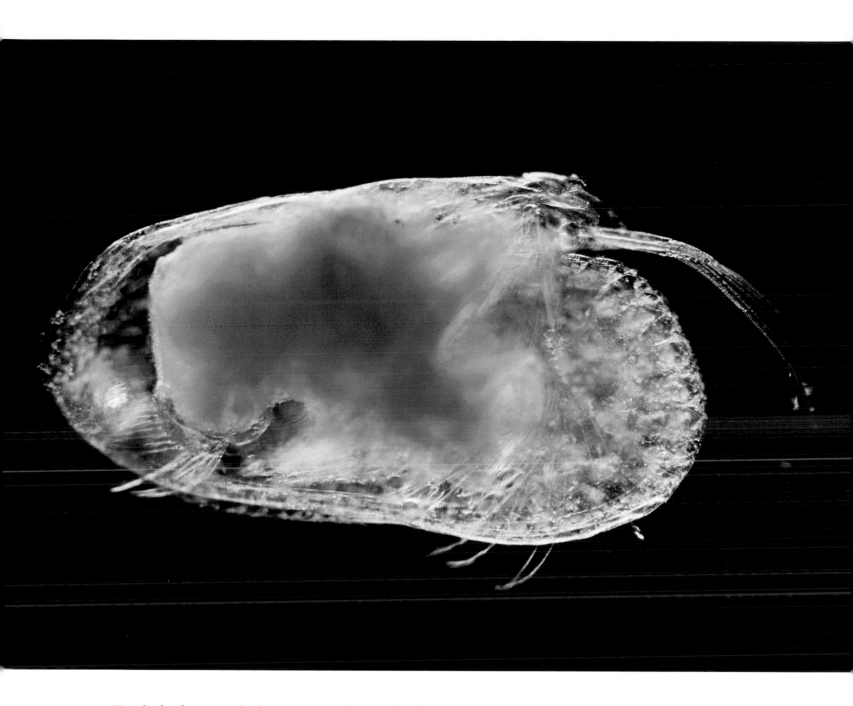

Hundreds of new zooplankton species are being identified by Census researchers, who expect to double the 7,000 known species within the next decade. This potential new ostracod (Archiconchoecetta sp.) was collected from the southern Atlantic Ocean off the coast of Namibia.

terms run the gamut from a physical attribute of the species (such as hairy legs) to its geographic location to a word chosen to honor someone or something. One of the first species named by Census researchers studying the Mid-Atlantic Ridge, for example, was the squid *Promaco-teuthis sloani*. Its species name recognizes the contribution of the Alfred P. Sloan Foundation, which provided support for the initial launch of the Census project.

The final step in naming a new species is to submit the written description to a peer-reviewed scientific journal. Only when the article has been accepted and published can the species be officially designated a new member of Earth's biological community.

IDENTIFYING THE DRIFTERS

While the typical process of species identification is laborious, for some scientists the Census made their labors much more exciting and efficient. Census taxonomists who study zooplankton, for example, were given a unique opportunity to take the identification process to sea. Zooplankton are marine animals that drift in the water rather than swim, so they depend upon the vagaries of ocean currents. They typically range in size from 1 millimeter to a few centimeters, although the nearly neutrally buoyant gelatinous zooplankton can grow to many meters (yards) in length.

Census zooplankton taxonomists, who have spent their careers becoming familiar with the morphology of these tiny drifting marine animals, were able to see live specimens firsthand as they were brought on board in fine mesh nets. Ann Bucklin, director of the Marine Science Technology Center of the University of Connecticut at Avery Point and head of the Census's zooplankton project, explains: "Identifying a new species is perhaps the most exciting thing a taxonomist does, for making sure a species is indeed new is a tough, exacting game for zooplankton. There's also an enormous sense of excitement looking at samples fresh, alive, that no one has ever seen before."

Bucklin predicts that Census taxonomists will at least double the number of known zooplankton species, which in 2008 consisted of 7,000 species in 15 different phyla. "Most of the zooplankton species are rare, so when a seagoing taxonomist samples an unexplored area such as 100 meters (325 feet) off the bottom in 5,000 meters (16,000 feet) of water, where it is now possible to sample, we find lots of new species." Like their larger counterparts, zooplankton are being discovered much

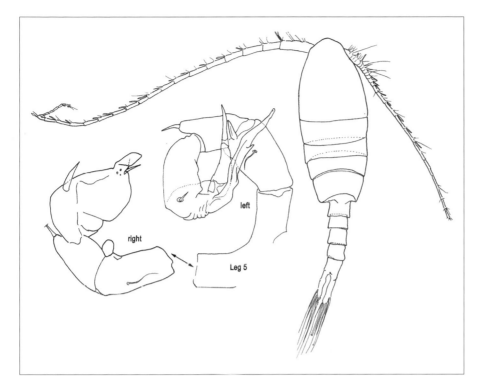

right

left

Leg 5

faster than they can be formally identified. To accelerate the process, zooplankton researchers are moving toward DNA barcoding to aid in species identification, which may prove particularly useful for zooplankton because the organisms are frequently rare, fragile and/or small.

Until recently, zooplankton taxonomists relied solely on illustrations to compare potential new species with identified species. Even today, careful drawings of new species are important for depicting difficult-to-observe details that are necessary to distinguish one species from another. For example, the structures in the drawing above are the legs of a male, and they are different in very subtle ways from other male legs of this genus. Often, critical examination of the spines and bumps on legs or mouthparts is the only way to separate one species from another. That is why DNA analysis has been such a boon to taxonomy — it can help answer questions such as whether that bigger bump or longer spine really indicates a different species. Genetic information makes the decision much easier, and possibly more exact.

FINDING A NEW OLD SPECIES

Census researchers identified a living species of "Jurassic" shrimp of the new genus Laurentaeglyphea, *which was thought to have become extinct nearly 50 million years ago.*

As the contents of a deepwater trawl from the Coral Sea were dropped on deck, one specimen stood out. A small straw-colored shrimp caught the eye of veteran researcher Bertrand Richer de Forges, who had seen his fair share of trawl contents over his 30-plus years as a marine scientist. Richer de Forges couldn't believe his eyes. On first examination, this intriguing little shrimp appeared to be a living specimen from a group of crustaceans that were thought to be extinct. Spotting it among the many specimens on deck started a chain of events that made biological history as the tale of the "Jurassic shrimp" unfolded.

Glypheid shrimp are a group of marine arthropods (invertebrates) that were abundant during the Jurassic period, 213 million to 144 million years ago.

They were believed to have become extinct during the Eocene period (about 50 million years ago) – until one was caught in 1906 off the coast of the Philippines by the U.S. research vessel *Albatross*. That specimen was preserved and tucked away in a museum collection for more than 60 years, until it was rediscovered in 1975 by two French scientists, who identified it as a glypheid and named it *Neoglyphea inopinata*. After three new deep-sea cruises in the area of the Philippines, 13 additional specimens have been caught, confirming that, while extremely rare, these glypheid shrimp are not extinct.

Enter the strange shrimp taken from the Coral Sea in 2005. The RV *Alis* had been trawling at a depth of 400 meters (1,320 feet) along a chain of seamounts south of the Chesterfield Islands

Bertrand Richer de Forges, Institut de recherche pour le développement, New Caledonia.

(between New Caledonia and Australia) when the living fossil was brought aboard in the nets. The crustacean appeared to be a glypheid, and while it superficially resembled *N. inopinata*, it was notably different in coloration and markings and substantially more robust than its predecessors. Suspecting that it was another living example of glypheid,

Richer de Forges consulted with colleagues and concluded that his was indeed a new species.

Richer de Forges then embarked on the process of convincing the larger scientific community about his conclusions. Studies of the specimen's morphology and analysis of known decapod taxonomy made the new species' position on the tree of life increasingly clear. Additional clues revealed more about the animal. For instance, the well-developed eyes and powerful pseudochelae (pincer-like appendages) suggested that the shrimp was a predatory animal. The location of its capture — a hard, rocky bottom — also differentiated it from *N. inopinata*, which is known to inhabit burrows in muddy bottoms.

A description of this new species, named *Neoglyphea neocaledonica*, was published in 2006. Professor Jacques Forest, one of the two scientists who first described *N. inopinata* in 1975, came to question whether this animal did belong to the same genus as its northern cousin, as its new name suggested. In the end he decided to create a new genus of glypheid for this species, and named it for his colleague, Michele de Saint Laurent. Richer de Forges' manuscript was eventually published in a peer-reviewed journal, and a new species (with very old ancestors), *Laurentaeglyphea neocaledonica*, was introduced to the world.

Census taxonomists are not only finding new species but also documenting many other firsts – new knowledge about previously known species. One example is this deepwater copepod (Eaugaptilis hyperboreus); *Census scientists were the first to photograph one bearing its eggs.*

WHERE NO ONE HAS GONE BEFORE

Seven years into its 10-year effort, the Census of Marine Life had discovered more than 5,300 new species. By 2010 it is estimated that the number of new species discovered by Census scientists will have increased to around 10,000, causing the number of known, named marine species to grow to approximately 240,000. While this is noteworthy, many more discoveries are yet to be made. Senior scientist Ron O'Dor estimates that when this first Census is complete, anywhere from 200,000 to 2 million *more* marine species will remain to be discovered, described and named.

"Of course, we're finding new species because we're going to places that haven't been explored before. It's exciting – humbling, really – to be the first to see a fish or other marine critter that no one else has ever seen before," says Richard Pyle, a Census scientist at Bishop Museum in Honolulu, who has spearheaded the use of sophisticated closed-circuit rebreathers, which recycle breathing gas – helium, nitrogen and oxygen – to allow divers to descend to previously off-limits depths. Pyle and his scientific team have found more than a hundred new species of coral-reef fishes. As many as 28 new species were discovered on one diving expedition alone, in April 2007 to the deep coral reefs across the Caroline Islands in the Pacific Ocean.

The Caroline Islands expedition, sponsored by the British Broadcasting Corporation (BBC), was filmed for a documentary about the discovery of new species of deepwater coral-reef fishes. Of the team's many new finds, the most conspicuous and unambiguously new species were several in the damselfish genus *Chromis*. The most spectacular of these was an intensely blue fish that lives at 120 meters (400 feet) and below. In keeping with the traditions of naming new species, the scientists named the fish *Chromis abyssus*, in recognition of its color and its deep habitat (relative to most species of the genus) and to honor the documentary film project – *Pacific Abyss* – that supported the expedition.

The plethora of new species discovered during just one of Pyle's expeditions demonstrates the benefits of venturing into previously unexplored areas. Much like the imaginary *Star Trek* crew, the Census's credo is to go where no human has gone before, which has presented some scientific challenges. By exploring unexplored territories and investigating everything from microbes to whales, Census and other explorers are finding unknown species faster than they can be described – not enough

This striking blue damselfish was named Chromis abyssus *in recognition of the BBC's support of its discovery, as well as for its outstanding color and deepwater home.*

experts are available to do the work. Philippe Bouchet, of the Muséum national d'Histoire naturelle in Paris, is a Census researcher with the continental margins project. He calculates that 3,800 taxonomists enter at least 1,400 new marine species into the literature every year. At that rate, the process of discovering, verifying, describing and naming all the remaining unknown marine species could take over five *centuries*. The fruits of Census ventures will continue to be realized for decades to come!

A WINDOW BELOW THE ICE

The quantity of new species discovered depends in large part on the region being explored. Exploration of the ice oceans – the Arctic Ocean in the Northern Hemisphere and the wild Southern Ocean at the other pole – has led to many interesting surprises about what lives in these cold, hostile regions. The research required dedication, perseverance and the right equipment. Ships to break through the ice to gain passage, sharpshooters to protect divers from roaming polar bears in the north, and the stamina to withstand violent seas in the Southern Ocean were just a few of the prerequisites for exploring these regions at the opposite ends of the earth.

In the deep Arctic Ocean – described by Census researcher Bodil Bluhm as "the most understudied area of the ocean world" – a team of daredevil explorers spent 30 days aboard the U.S. Coast Guard cutter *Healy* in an attempt to better understand what inhabits the depths of this cold and rapidly changing environment. Over that period they discovered 35 previously undescribed species that are now in the lineup for official designation. Bluhm's colleague Russ Hopcroft explains: "Modern technology opened a window on this amazing world for the first time. The imagery obtained at the mid-water and seafloor shows many life-forms such as soft-bodied zooplankton, deep-sea cucumbers and soft corals. The few explorers in this area before us had no adequate tools to collect or see these creatures."

In terms of species numbers, the wealth of discoveries in the Southern Hemisphere, where even seasoned explorers were surprised to find such high levels of biodiversity, far exceed those in the north. Three Census expeditions to the Weddell Sea, a deep embayment of the Antarctic coastline, were launched between 2002 and 2005 to fill the

Below left: A potential new species of comb jelly, a cydippid ctenophore, was discovered by Census scientists during a month-long expedition to the Arctic Ocean in 2005.

Below right: One of three newly described species found during a Census expedition to the Canada Basin in the summer of 2005, this polychaete worm (genus Macrochaeta*) was found on the floor of the Arctic Ocean.*

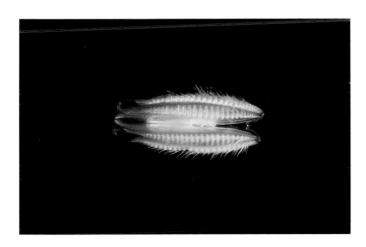

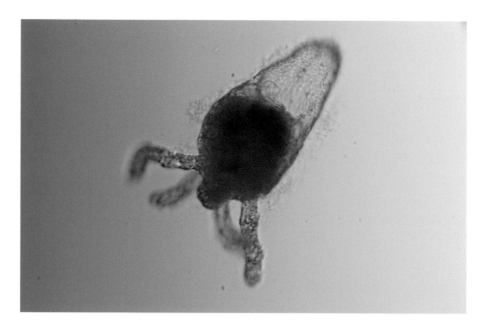

A presumed new species of Narcomedusae, *a type of jellyfish, was first collected in 2002 south of Banks Island in the Canadian Arctic, and then collected and photographed again in 2005.*

"knowledge vacuum" surrounding the fauna of the deeper parts of the Southern Ocean. In these deep southern regions, three international teams collected tens of thousands of specimens from water depths of between 774 and 6,348 meters (approximately 0.5 to 4 miles).

Once the samples were brought home, researchers wholeheartedly undertook the challenge of distinguishing the new from the old. Much to their delight, the scientists had found more than 700 potentially new species, 6 of which were carnivorous sponges. Included in the catch they spotted 674 species of isopod (a diverse order of crustaceans), most of which had never been previously described; more than 200 polychaete species (marine worms), 81 of which were new; and 76 sponges, 17 of which had previously been unknown.

Expedition leader Angelika Brandt suggests that the Antarctic deep sea may be the cradle of life of global marine species and that the findings could shed light on the evolution of ocean life in this area and beyond. By comparing species found in the deep sea and those found in the shallower waters surrounding Antarctica, scientists may be able to better understand how climate and the environment in which these animals live have driven past evolutionary changes, and could portend how they may be able to adapt in the future.

As part of the enhanced exploration efforts of the 2007–09 International Polar Year (IPY), the Census project that focused on the Antarctic – the Census of Antarctic Marine Life (CAML) – spearheaded 18 expeditions to the Southern Ocean. The first of these, which returned in

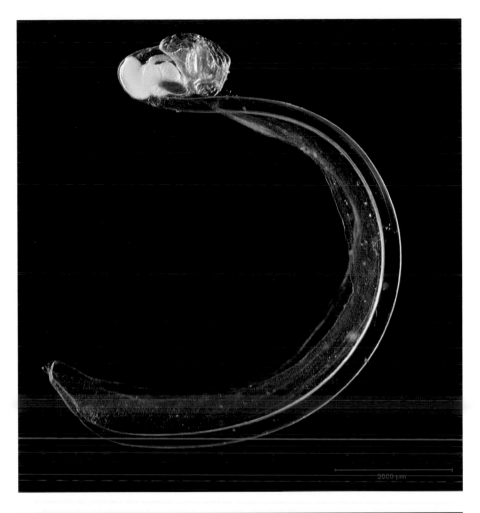

This potential new larvacean species was discovered in the Arctic's Canada Basin.

This Munnopsis *species, looking like a deep-sea spider, was found in the western Weddell Sea. It is a type of isopod – a group of marine invertebrates (animals without backbones) that eat bits of food that sink to the seafloor.*

March 2007, coincided with the launch of the IPY, and it showcased the opportunities for increasing understanding of this complex ecosystem.

In November 2006, a team of 52 scientists from 14 countries began a 10-week expedition aboard the German research vessel *Polarstern* to explore an area no one had ever studied before. The scientists investigated a 10,000-square-kilometer (4,000-square-mile) portion of the Antarctic seabed that had only recently become accessible to exploration. This opportunity was made possible by the collapse of the Larsen A and B ice shelves, which together comprised an area about the size of Jamaica.

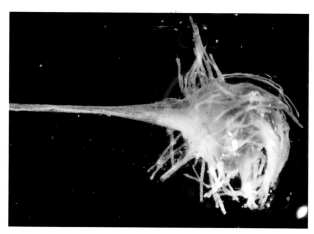

Asbestopluma, *a never-before-known species of carnivorous sponge about 1 centimeter (0.4 inch) in diameter, engulfs organisms and then digests the imprisoned prey. This is one of four such species, three of them believed new to science, found in the Southern Ocean abyss.*

With sophisticated sampling and observation gear, including a camera-equipped remotely operated vehicle, experts on the *Polarstern* returned with revealing photography of life on the seabed that had been uncovered by the disintegration of Larsen A and B. Among an estimated 1,000 species collected, several are proving to be new to science.

"This knowledge of biodiversity is fundamental to understanding ecosystem functioning," says Julian Gutt, a marine ecologist at Germany's Alfred Wegener Institute for Polar and Marine Research and chief scientist on the *Polarstern* expedition. "The results of our efforts will advance our ability to predict the future of our biosphere in a changing environment."

Many hundreds of animal specimens were collected during the voyage, including 15 potentially new amphipod (shrimp-like) species. The star attraction was one of Antarctica's largest amphipod crustaceans; nearly 10 centimeters (4 inches) long, it is larger than many similar species found in temperate climates. Four presumed new species of cnidarians – organisms related to coral, jellyfish and sea anemones – were also collected.

Left: The Larsen B ice shelf as it appeared in January 2002.

Right: The area after the ice shelf collapsed in March 2002.

 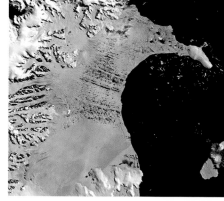

Left: A presumed new species of Epimeria, *a 25-millimeter-long (almost an inch) amphipod crustacean, was collected near Elephant Island during an expedition to the Weddell Sea in 2006–07.*

Below: This potentially new giant Antarctic amphipod crustacean – of the genus Eusirus – *was one of the stars among the species collected during the trip to the Weddell Sea in 2006–07. Nearly 10 centimeters (4 inches) long, it was sampled by using baited traps off the Antarctic Peninsula.*

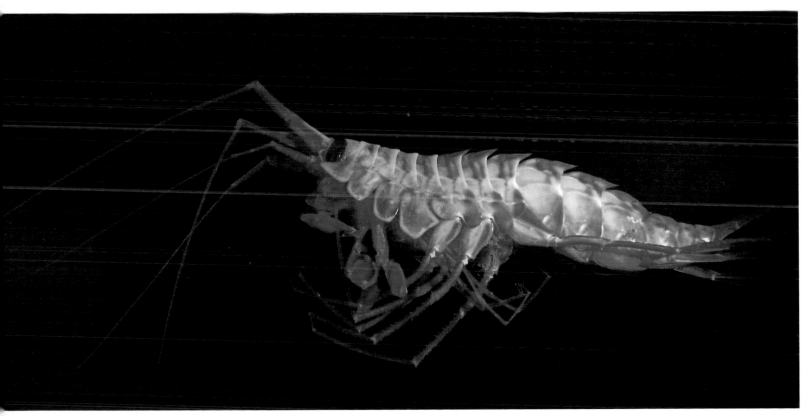

A well-known species of giant barnacle, Bathylasma corolliforme, *was photographed during the* Polarstern *expedition to the Southern Ocean in 2006, where many unknown species were collected and photographed for future identification.*

This unidentified Antarctic sea star was sampled from the area formerly covered by the Larsen B ice shelf in the Weddell Sea.

ON BEING THE FIRST TO FIND
A NEW LIFE-FORM

Steven Haddock, of the Monterey Bay Aquarium Research Institute, is the author of this essay.

Discovering a species generates a lot of different reactions. First is definitely amazement that here is the embodiment of yet another variation on the shapes, sizes and structures within the basic pattern of what it takes to be a jellyfish, for example. Working in the deep sea, we find things that can be radically different from known species, not just by an eyelash. Sometimes the reaction is skepticism: someone must have seen this before. The feeling must be akin to finding a chest of buried treasure.

After the initial excitement comes puzzlement about how this relates to species that you already know. There is some kind of natural human instinct to categorize things. My favorites are the critters that we have no clue what to temporarily call them. Such animals are given descriptive pet names

like "Mystery Beast," "Weird Ctenophore," "Blue Siphonophore," "Green Bomber" and so on. These names are useful (in moderation) for referring to the unknowns when we come across them again, and also so we can get a toehold on making sense of them.

It is almost impossible to make a living doing only taxonomy, that is, describing specimens. Species descriptions require a lot of painstaking work, literature searches and very dry scientific writing, but they are a necessary task of completing science that takes a broader view. If you want to discuss the ecological niche of an organism, clone its genes or place it in the broader community context, you need a name for it (you can go only so far with "Blue Siph").

So eventually the excitement can be seasoned with a bit of frustration – "here's another year of work added to the pile!" But as the process unfolds, I never cease to be amazed by the tremendous amount of diversity that lies just beneath the surface, waiting to be discovered.

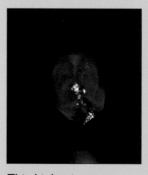

This bioluminescent siphonophore is a recently discovered carnivorous plankton notable for its dramatic blue coloration. It was named Gymnopraia lapislazula after lapis lazuli, a flecked, intensely blue stone. Lapis, Latin for "stone" or "milestone," was also chosen to commemorate the 25th siphonophore description by Philip R. Pugh, one of the descriptive paper's authors.

EYES AND EARS IN THE DEEP

Outside of the polar regions, the rest of the deep, dark zone of the ocean – from 200 meters to more than 5,000 meters (660–16,500 feet) deep – has also contributed its fair share of new species. The impediments of depth, pressure and inaccessibility had to be overcome to increase understanding of what lives "deeper than light" in the global ocean. The discoveries were made possible by deploying nets and trawls that can sample to great depths and by new technology – such as remotely operated vehicles equipped with high-definition cameras and video equipment, as well as autonomous submarines – that can sample under the direction of operators who remain safely on the surface. This equipment has opened up new windows to what lives below the surface.

For example, Census expeditions to the Mid-Atlantic Ridge, an underwater mountain range in the middle of the Atlantic Ocean, collected tens of thousands of organisms by using sophisticated equipment that included the Russian manned submersible *Mir*. Scientists believe they found around 30 new species, but by the end of 2008 only 5 had been officially designated as such – another testament to the rigors of the identification process.

Odd Aksel Bergstad, of the Institute of Marine Research in Arendal, Norway, who led the Census project focused on the Mid-Atlantic Ridge, says, "Because we study macro- and megafauna only – rather big animals that are expected to be well studied and less diverse than smaller organisms or those found in very species-rich places such as highly diverse coral reefs – we didn't really expect to come across new species at all. Hence we were pretty excited to be able to contribute to the growing number of marine species."

Other Census expeditions to the deep – to study hydrothermal vents and cold seeps and continental margins – have also contributed additions to the marine species family tree.

Below left: This new species of eelpout of the genus Lycodonus *was collected on the Mid-Atlantic Ridge.*

Below right: This new species of grenadier (a.k.a. rattail), Caelorinchus mediterraneus, *was found in the western Mediterranean.*

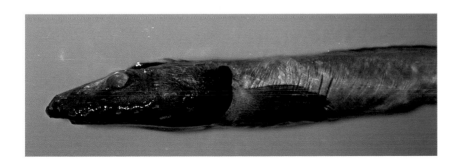

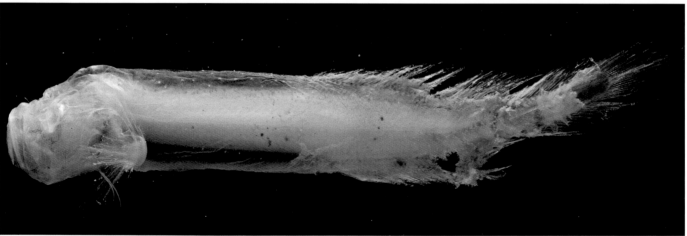

Top: This new species of crab was discovered off the Easter Island Microplate. It was named Kiwa hirsuta kiwa *after the goddess of shellfish in Polynesian mythology, but has become known as the "yeti crab" because of its hairy appearance.*

Bottom: The Aphyonidae are a family of fish that live below 700 meters (2,300 feet), and all species are very rare. This gelatinous specimen of the genus Barathronus *is most likely an undescribed species.*

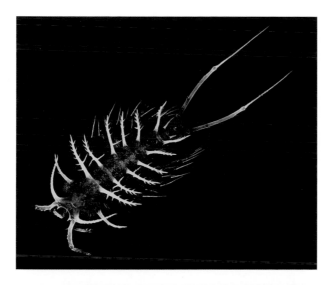

Top: Census scientists are finding copepods, the smallest deep-sea animals, in places they never expected. This bizarre new copepod (Ceratonotus steiningeri) *was first discovered 5,400 meters (18,000 feet) deep in the Angola Basin in 2006. Within a year it was also collected in the southeastern Atlantic, as well as some 13,000 kilometers (8,000 miles) away in the central Pacific Ocean. Scientists are as puzzled about how this tiny (0.5 millimeter) animal achieved such widespread distribution as they are about how it avoided detection for so long.*

Bottom: Researchers exploring the Mississippi Canyon, which extends from the Louisiana shore into the Gulf of Mexico, found a carpet of crustaceans (Ampelisca mississippiana).

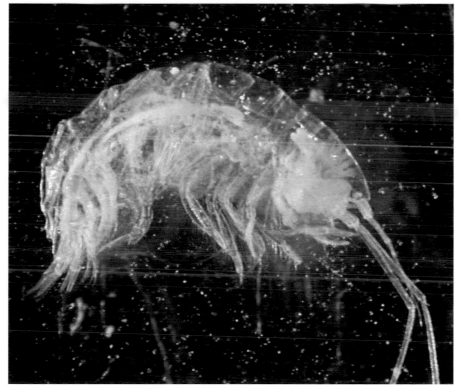

Technological advances have made it possible to identify new life-forms that only a few years ago were virtually impossible to see. A revolutionary new DNA technique, 454 tag sequencing, requires only small snippets of genetic code to identify an organism. With its help, Census scientists have revealed that the diversity of marine microbes may be some 10 to 100 times greater than expected. By far the greatest part of this unexpected diversity is previously unknown, low-abundance organisms believed to play an important role in the marine environment as part of a "rare biosphere."

Marine microbes are so plentiful in this image that they resemble stars in a night sky. Census scientists found 20,000 different kinds of bacteria in one liter of seawater, many previously unknown and scarce.

"These observations blew away all previous estimates of bacterial diversity in the ocean," says Census scientist Mitchell L. Sogin, director of the Marine Biological Laboratory's Josephine Bay Paul Center for Comparative and Molecular Biology and Evolution, at Woods Hole, Massachusetts, and leader of the Census project investigating microbes. "Just as scientists have discovered through ever more powerful telescopes that stars number in the billions, we are learning through DNA technologies that the number of marine organisms invisible to the eye exceeds all expectations, and their diversity is much greater than we could have imagined." Before this study, microbiologists had formally described 5,000 microbial "species," but using this new DNA technique, scientists discovered more than 20,000 in a single liter (about a quart) of seawater.

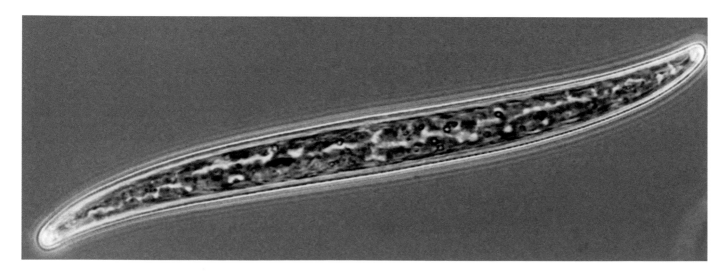

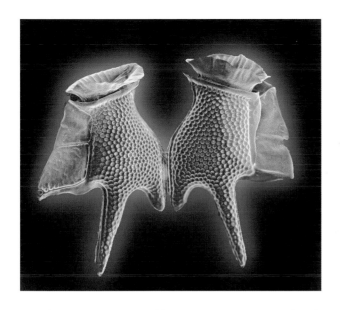

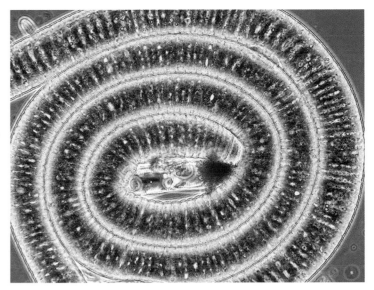

Modern technology is allowing scientists to see the abundance and variability of microbial life-forms that come in a variety of colors, shapes, sizes and structures, and in heretofore unimaginable quantities.

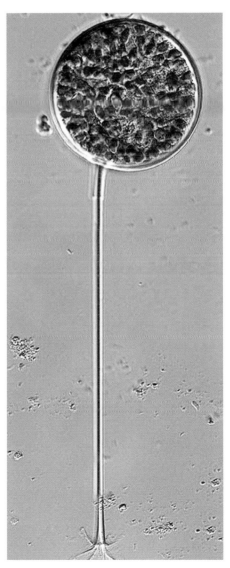

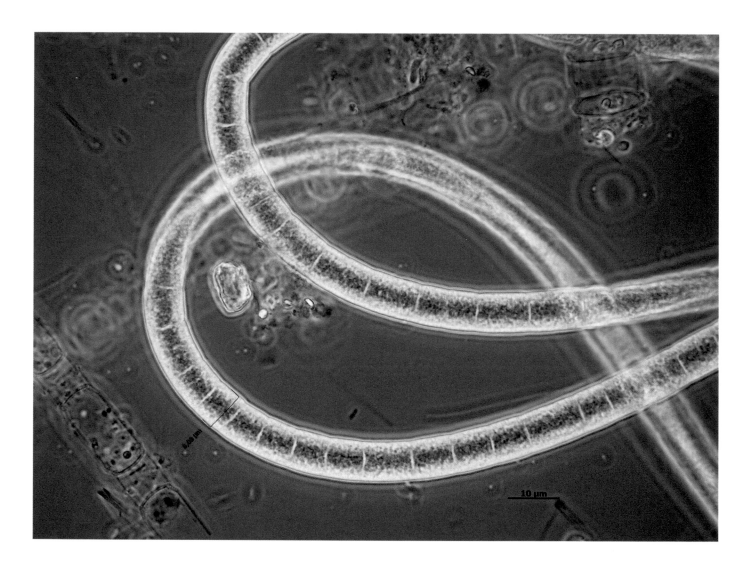

10 µm

A presumed new species of giant sulfur bacteria found in the southeast Pacific Ocean may provide insight into early forms of life on Earth, and could provide a potential model in the search for extraterrestrial life.

LOOKING AHEAD

The coral reefs project was a latecomer to the Census of Marine Life, becoming operational only in 2005. Yet, in a remarkably short period, it has made considerable progress. During one three-week expedition to French Frigate Shoals in 2006, for example, 50 sites were sampled and 2,500 samples collected. Of these, scientists project that at least 100 new species are likely, with the potential for several new species of crabs, corals, sea cucumbers, sea squirts, worms, sea stars, snails and clams to follow.

Another Census sampling effort was conducted during September–October 2006 to document the fauna and flora of the remote South Pacific island of Espiritu Santo in Vanuatu. The number of species inventoried there was similarly remarkable: 1,100 species of decapod crustaceans (crabs, shrimps and hermit crabs) and about 4,000 species of mollusks. Hundreds of species – possibly more than a thousand – were believed to be new to science.

In 2008, Census scientists began to systematically explore waters off two islands on the Great Barrier Reef and a reef off northwestern Australia – waters long familiar to divers – and were surprised to find hundreds of potentially new animal species. The expeditions marked the International Year of the Reef and included the first systematic scientific inventory of spectacular soft corals, named octocorals for the eight tentacles that fringe each polyp.

Discoveries at Lizard and Heron Islands (part of the Great Barrier Reef), and Ningaloo Reef in northwestern Australia, were many. Scientists collected about three hundred soft coral species, up to half of them thought to be new to science, along with dozens of small crustacean species – and potentially one or more *families* of species – also thought new to science. They also gathered new species of tanaid crustaceans (shrimp-like animals, some with claws longer than their bodies), along with scores of tiny amphipod crustaceans (insects of the marine world) of which an estimated 40 to 60 percent will be formally described for the first time. As well, the researchers deployed new methods designed to enhance comparison of coral reefs worldwide by standardizing measurement of reef health, diversity and biological makeup.

These were but the first advances in knowledge about thousands of species at risk in coral reef environments around the globe. Undoubtedly much more will be learned as additional coral reef surveys are undertaken through 2010.

PART THREE

WHAT WILL LIVE IN THE OCEAN?

CHAPTER NINE

Forecasting the Future

Humans have always been very good at killing big animals. Ten thousand years ago, with just some pointed sticks, humans managed to wipe out the woolly mammoth, saber tooth tigers, mastodons and giant vampire bats. The same is now happening in the oceans.

– The late Ransom A. Myers, fisheries biologist and head of the Future of Marine Animal Populations Project

Earth is undergoing rapid climatic warming. The risk is high that projected climate changes will have a severe impact on biodiversity, which would exacerbate other pressures (such as habitat loss) and lead to substantial increases in species extinction and ecosystem disruption. The common impression is that marine species and ecosystems are generally in good shape. As more is learned, however, that impression is turning out to be false. When populations of a marine species become depleted, genetic variation is reduced, which compromises the species' ability to adapt to new environmental changes and stresses such as global climate change. Because species depend upon each other for their survival, the demise of one can lead to the decrease or demise of others. When species disappear or remain only few in number, ecosystems may not possess enough species and genetic diversity to enable them to survive.

The Census is describing and synthesizing globally changing patterns of species abundance, distribution and diversity and modeling the effects of fishing, climate change and other key variables on those patterns. This work, done across ocean realms, emphasizes understanding

Pages 202–203: The football fish (Himantolophus paucifilosus) *is dotted with pearl-like stitches, each one of them a sensory organ. These organs protrude through the skin's surface and allow the fish to detect even a slight displacement of water.*

Opposite: Guadalupe fur seals (Arctocephalus townsendi) *were once thought to be so scarce that extinction might be unavoidable. A sizeable population can now be found at Guadalupe Island, 240 kilometers (150 miles) off the Mexican mainland.*

Above: This is just a small portion of the huge school of cownose rays (Rhinoptera steindachneri) *that circled a dive site in the Galapagos Islands.*

Opposite: Underwater encounters with endangered Hawaiian monk seals (Monachus schauinslandi) *are few and far between – the population is estimated at only slightly more than 1,000 individuals. Even scarcer is the Mediterranean monk seal, with a population of less than 500. The Caribbean or West Indian monk seal* (Monachus tropicalis), *the only seal ever known to be native to the Caribbean Sea and Gulf of Mexico, was last seen in 1952. In 2008, after five years of trying to find evidence of Caribbean monk seals, the U.S. government declared the species officially extinct.*

past changes and predicting future scenarios. Scientists are aware that many known species are under threat of extinction, which has major implications for the loss of the rich diversity necessary for healthy ecosystems. The Census is racing the clock to identify unknown species and to ascertain the risks to marine life.

An extinction event is a sharp decrease in the number of species in a relatively short period. A global mass extinction event affects most major taxonomic groups – birds, mammals, reptiles, amphibians, fish, invertebrates and other, simpler life forms. These events affect the earth as a whole and may occur in a relatively short period of geologic time. Marine fossils have been used to measure global extinction rates because they are more plentiful than terrestrial fossils.

Since life began on Earth, several major mass extinction events are believed to have taken place. Fossil evidence from the past 550 million years provides evidence of five of these events, when more than 50 percent of living animal species were lost. The most recent event was at the Cretaceous/Tertiary (K-T) boundary, approximately 65 million years ago, when the dinosaurs died out. The causes of mass global extinctions are not thoroughly understood, but abrupt climatic changes may be the initiating factor. In the case of the K-T event, many scientists now believe that massive volcanic activity may have reduced sunlight, cooling

Earth and disrupting life across the globe. In addition, a huge asteroid impact may have accelerated extinctions. Of the five mass extinction events, four – including the one that eliminated the dinosaurs – are associated with greenhouse phases. The largest extinction event of all, at the Permian/Triassic boundary 251 million years ago, occurred during one of the warmest climatic phases ever in Earth's history; it saw the estimated extinction of about 95 percent of all species!

Scientists believe that Earth is now in the midst of a sixth global mass extinction event, and biologists theorize that this event may be one of the largest and fastest ever to occur. Some scientists, such as E.O. Wilson of Harvard University, predict that degradation of habitats caused by human activities combined with rapid climatic changes could cause the extinction of half of all remaining species in the next 100 years. According to a United Nations report on the state of the global environment, 25 percent of all mammal species are expected to become extinct in the next 30 years.

The above statistics refer to all life on Earth. Census scientists paint an even bleaker picture for the ocean's fish. A global Census study concludes that 90 percent of all large fishes have disappeared from the world's ocean in the past half-century, the devastating result of industrial

Orange roughy (Hoplostethus atlanticus) *is a relatively large deep-sea fish. Slow-growing and late to mature, it may attain a long life – a recorded maximum of 149 years. It is thus extremely susceptible to over-fishing, and many stocks, especially off New Zealand and Australia, have already crashed.*

fishing. This research, published by the late Ransom Myers, is one of the Census's foundational studies. The investigation used data going back 47 years, from nine oceanic and four continental shelf systems ranging from the tropics to the Antarctic.

"I think the point is – there is nowhere left in the ocean not overfished," said Myers, a fisheries biologist. He reported that the big declines in the large fishes began when industrial fishing started in the early 1950s. "Whether it is yellowfin tuna in the tropics, bluefin in cold waters, or albacore tuna in between, the pattern is always the same. There is a rapid decline of fish numbers," Myers said.

The pressures of fishing have given rise to a new category of species depletion: commercial extinction, in which fish and shellfish populations are depleted to the point where it is no longer economically feasible to fish for them. While not extinct, these species certainly no longer play their traditional role in their ecosystem; some, such as white abalone off the coast of California, have been pushed to the brink of extinction. Fishing operations such as trawling and dragging destroy bottom habitats and deplete species populations. The continuation of these operations delays or prevents recovery.

Ransom Myers's research partner, Boris Worm, a professor at Canada's Dalhousie University, says that the losses are having a major impact on ocean ecosystems. Predatory fish are like "the lions and tigers of the sea," declares Worm. "The changes that will occur due to the decline of these species are hard to predict and difficult to understand. However, they will occur on a global scale, and I think this is the real reason for concern." After the 2003 Census study reported that 90 percent of the top predator fish had vanished from the oceans, Worm stated, "This problem is silent and invisible. People don't imagine this. It hasn't captured our imagination, like the rainforest."

In many cases, fish numbers plummeted fastest during the first years after fleets moved into new areas, often before anyone knew drops were taking place. Longlining, among the most widespread of fishing methods, uses miles of baited hooks to catch a wide range of species. Fifty years ago, fishers using longlines would catch about 10 big fish per 100 hooks. Now, according to 2003 Census reports, the norm is one fish per 100, and the fish are about half the weight of earlier years. In 2003 Myers warned that the world's great fish could go the way of the dinosaurs – in large part as a result of longlining – if immediate action was not taken.

Other species besides fish are in danger from longlines. Loggerhead and leatherback turtles have an annual 40 to 60 percent chance of meeting a longline hook, and thousands are dying as a result, says Census scientist Larry Crowder. The Duke University researcher reported in 2004 that urgent action was needed to prevent these creatures from disappearing from the ocean. "In the year 2000, longline fishermen from 40 nations set at least 1.4 billion hooks on longlines that average about 40 miles long," he said. "That's 3.8 million hooks per night that are set globally."

Boris Worm says he hopes that the Census's "big picture" study of the world's fish populations will serve as a wakeup call to governments, global fishing conglomerates and environmental groups. "People haven't before seen how bad this is. It doesn't make any sense, economically or ecologically, to ignore this."

While the numbers are alarming, Worm says that there are solutions. In the past when certain fishing areas were declared off limits and restrictions were enforced, certain fish and shellfish populations rebounded "amazingly quickly." Haddock, yellowtail and scallops have recovered in different regions. However, with numbers down so dramatically in every part of the world, the situation cannot be ignored. Census scientists estimate that reversing the decline in global fisheries would require cutting back global fishing by as much as 60 percent.

This bluefin tuna weighing 1,305 kilograms (593 pounds) was caught in the 1971 United States Atlantic Tuna Tournament off Point Judith, Rhode Island.

Right: Sound conservation and management offer hope that longline fishing can be sustainable for certain species. These 1991 fishermen in the Chatham Strait, a narrow passage off southeastern Alaska, are hauling in a 1.2-kilometer (0.75-mile) longline with hooks every 4.5 to 6 meters (15–20 feet). Note the large sablefish (Anoplopoma fimbria), *also called black cod, in the water. More than 80 percent of this long-lived species – the maximum recorded age is 94 years – are landed using longline fishing. This fishery is managed by the North Pacific Fishery Management Council, which has established annual individual sablefish quotas.*

Below: Endangered humpback whales (Megaptera novaeangliae) *can be heard "singing" in the waters off Maui, Hawaii. Their eyes, about as large as grapefruit, are located behind their mouth.*

Fish and turtles are not the only species at risk; marine mammals are also under great duress. Species on the endangered or threatened species list include southern sea otters, manatees, Guadalupe fur seals, monk seals and humpback, blue, fin, sei, right and bowhead whales. The western Pacific population of the Steller sea lion was added to the endangered list in 1997 in response to an 80 percent drop in its population over the past 30 years, while the eastern population is still listed as threatened.

Above: The remaining endangered Steller sea lions (Eumetopias jubatus) *are found in the North Pacific. Among pinnipeds, they are smaller than only the walrus and the elephant seals. The Steller sea lion has attracted considerable attention in recent decades because of significant unexplained declines in their numbers over a large portion of their range in Alaska.*

California sea otters (Enhydra lutris), *whose numbers were once estimated at 150,000 to 300,000, were hunted extensively for their fur between 1741 and 1911, reducing the world population to 1,000 to 2,000 individuals in a fraction of their historic range. A subsequent international ban on hunting, conservation efforts and reintroduction programs in previously populated areas have contributed to a rebound in the population. The recovery of the sea otter is considered an important success of marine conservation, although populations in the Aleutian Islands and California have recently declined or plateaued at depressed levels. It continues to be classified as an endangered species.*

It is difficult, and in many cases impossible, to determine the status of most species in the ocean. So little is known of many species' distributions or ranges that it cannot be determined whether they are plentiful or naturally rare, whether populations are stable or changing, and if they are threatened or endangered. Marine species that are relatively easily monitored are restricted to nearshore habitats, especially if they are sedentary or attached (such as sea grasses and corals), or spend time at the sea surface or on land (for example, marine mammals and seabirds). Census research has contributed a great deal of information about several marine species whose state was previously unknown. It is imperative that the research continue to fill in the remaining gaps in our understanding of marine biodiversity.

Through modeling and statistical work, the Census is working to determine what may live in the ocean of tomorrow. The large research questions being addressed include:

- What are the global patterns of marine biodiversity?
- Which are the major drivers that explain diversity patterns and changes?
- How many species are there in the ocean, both known and unknown?
- How has the abundance of major species groups changed over time?
- What are the ecosystem consequences of fishing and climate change?
- How are animal distributions in the ocean changing?
- How is the movement of animals determined by their behavior and environment?

The following case studies are examples of how attempting to address some of these important questions led to new understanding of how marine life is intertwined and how greatly ocean species rely on each other.

THE DEMISE OF THE GREAT SHARKS

Ecologists have long understood that a reduction in predators affects the entire food web, that complex network of interactions between plants and animals that tells us who is eating what or whom. Census researchers investigating this premise determined that loss of predator species from oceanic food webs can cause long-lasting changes in the ecosystem that may be irreversible. Researchers specifically studied the 11 species of sharks that scientists call the "great sharks," whose diets consist of other

Opposite: Cownose rays are flourishing now that their main predator species, the great sharks, have been depleted. The rays flap their fins rapidly and blow sand out of their gills to churn up sediment and expose hidden oysters, crabs and other shellfish. Powerful tooth plates snap and grind the shells, much like a nutcracker.

Scallops (Chlamys hastata)*, like this one off the coast of British Columbia, are one of the favorite foods of cownose rays. The small dots along the shell edges are the scallop's eyes.*

elasmobranchs (rays, skates and small sharks). Census research has shown that the populations of great sharks have been decimated, and the largest individuals of these top predators have also been lost, as indicated by declines in the mean length of blacktip, bull, dusky, sandbar and tiger sharks. These losses suggest that overexploitation has left few mature individuals in these populations.

With fewer predatory sharks in existence, there has been an explosion of their prey species in coastal northwest Atlantic ecosystems, and the effects of this community restructuring have affected the entire food web. For example, cownose rays have increased to the extent where one of their prey species, the bay scallop, has been so reduced in numbers that a century-long scallop fishery has come to an end.

The reduction of shark populations has caused a high level of international concern, and efforts are growing to conserve them. However, scientists are uncertain whether such initiatives might be too little, too late. The global demand for shark fins and meat has not ebbed. Shark fin

Although now rare and seldom seen, the sandbar shark (Carcharhinus plumbeus) is probably still the most numerous shark species found in Hawaii.

Scalloped hammerhead sharks (Sphyrna lewini) can still be found off Hawaii, but in limited numbers.

Bull sharks (Carcharhinus leucas) are known for their ability to survive in fresh water, where they give birth to their young. They are under great environmental stress because of their need for relatively pollution-free waters. This specimen was photographed in Bequ Lagoon, Fiji.

Dusky sharks (Carcharhinus obscurus) *have been overfished for their meat, oil and fins.*

As its dorsal fin slices the surface, the great white shark (Carcharodon carcharias) *is the perfect portrait of a predator. This specimen was photographed in the waters of South Australia.*

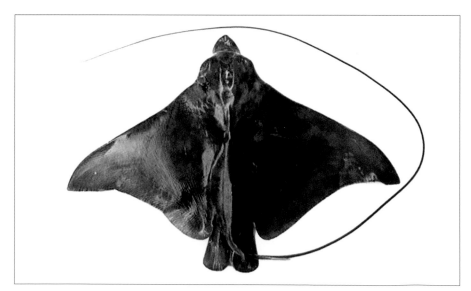

The longheaded eagle ray (Aetobatus flagellum) *population is increasing off the coast of Japan and decimating shellfish populations.*

Spotted eagle rays (Aetobatus narinari) *are becoming increasingly common off Hawaii.*

soup is a Chinese delicacy that is in high demand for weddings and other special occasions as a symbol of wealth and abundance. The practice of fishing sharks for their fins is controversial and problematic; it is considered a contributing factor in the global shark decline. Another problem for sharks is bycatch – when commercial fishermen catch sharks in the process of going after their species of choice. The sharks are thrown back overboard, most often dying or injured. An estimated 50 million sharks are unintentionally caught this way each year.

Census findings have been supported by the work of other researchers. The University of North Carolina has been conducting the Eastern Seaboard's longest continuous shark-targeted survey since 1972. The data demonstrate that sufficiently large declines in great sharks may lead to their elimination. Declines in seven species range from 87 percent for sandbar sharks (*Carcharhinus plumbeus*); 93 percent for blacktip sharks (*C. limbatus*); up to 97 percent for tiger sharks (*Galeocerdo cuvier*); 98 percent for scalloped hammerheads (*Sphyrna lewini*); and 99 percent or more for bull (*C. leucas*), dusky (*C. obscurus*) and smooth hammerhead (*S. zygaena*) sharks. Because this survey is located where it can intercept sharks on their seasonal migrations, these abundance trends may indicate Atlantic coast–wide population changes.

Census analyses show that shark prey populations may each have increased by about an order of magnitude. Overall, the research shows increases over the past 16 to 35 years for 12 of the 14 species. One of the largest gains has been made by the cownose ray (*Rhinoptera bonasus*).

This incredibly abundant population consumes large quantities of bivalves, particularly bay scallops (*Argopecten irradians*), soft-shell clams (*Mya arenaria*), hard clams (*Mercenaria mercenaria*), oysters (*Crassostrea virginica*) and several smaller, noncommercial bivalves. Cownose rays' annual bivalve demand just within Chesapeake Bay may approach 925,000 tons (840,000 metric tons) a year; in comparison, the 2003 commercial bivalve harvest in Virginia and Maryland totaled only 330 tons (300 metric tons). Increased predation by cownose rays may now be inhibiting recovery of hard clams, softshell clams and oysters, compounding the effects of overexploitation, disease, habitat destruction and pollution, which have depressed these species.

Similar ecosystem problems are occurring in other coastal regions. Studies in the northeast Atlantic Ocean have shown increasing abundances of several shark prey species. In Japan's Ariake Sound in the northwest Pacific Ocean, exploitation of predatory sharks is particularly intense. As a result, both wild stocks and cultured populations of several shellfish species are now being decimated annually by expanding numbers of longheaded eagle rays (*Aetobatus flagellum*).

CHANGES IN FISHERIES PRACTICE HELP WHALES

All is not completely bleak, however. Another Census study has shown how better management of the lobster fishery in Maine may benefit North Atlantic right whales, which remain critically endangered despite more than 70 years of protection. Their recovery is being hindered by accidental mortality caused by ship strikes and entanglement in fishing gear. Scientists studying photographs of right whales found that 75 percent bore evidence of entanglement, predominantly from lobster fishing gear. Lobster traps are tied to the surface via a buoy line and to other traps via ground lines, leading to entanglements as the whales swim and feed in the Gulf of Maine. Despite government regulations aimed at reducing entanglements, the problem is worsening.

The Census study demonstrates how Maine lobster fishers can protect the North Atlantic right whale without hurting their own bottom line. A team of researchers comparing the Nova Scotia and Maine lobster fisheries found that Maine lobster fishers could substantially reduce the number of traps and shorten the fishing season by as much as six months, and still catch the same amount of lobster, at lower cost. Doing

Immediate conservation and management are essential to ensure the future of innumerable ocean species, including the common bluestripe snapper (Lutjanus kasmira).

North Atlantic right whales (Eubalaena glacialis) *are critically endangered.*

so would protect right whales by reducing the risk of entanglements in fishing gear, a key obstacle to their recovery from the brink of extinction. "This is a classic win-win situation," says Boris Worm. "Given the high fuel and bait costs lobstermen incur, a shorter season and fewer traps will actually save money without reducing their catches."

This Census study also highlighted stark differences between the neighboring fisheries. Nova Scotia lobster fishers are restricted to a winter season, whereas in Maine they are permitted to fish year-round. In addition, Nova Scotians use 88 percent fewer traps than their Maine counterparts. Despite these differences, the actual landing patterns are similar – clear evidence of substantial wasted effort in the U.S. portion of the fishery. "Industry and government agencies have struggled to modify lobster and other fishing gear to reduce the risk of whale entanglement. But nothing can work as well as reducing the amount of excess gear in the water," says Census scientist Andrew Rosenberg of the University of New Hampshire.

Census research has been able to identify specific circumstances where changing or adapting management strategies can make a significant difference in the chances of endangered species or threatened ecosystems. One such study produced the first global assessment of the extent, effectiveness and gaps in coverage of coral reefs in Marine Protected Areas (MPAs). The Census team built a database of MPAs for 102 countries, including satellite imagery of reefs worldwide, and surveyed more than 1,000 MPA managers and scientists to determine the conservation performance of the protected areas.

Coral reefs are suffering around the world. Coral dieback and loss of reef biodiversity are now of such massive proportions that multinational efforts are being focused on reef conservation. The 2003 World Parks Congress recommended that 20 to 30 percent of all remaining coral reefs should be strictly managed by MPAs by 2012. At the time of their report (2006), Census scientists had determined that although 18.7 percent of tropical coral reefs are within MPAs, only 1.4 percent of these important ecosystems were inside "no-take" MPAs – areas with regulations covering extraction, poaching and other major threats. This research highlights the serious vulnerability of coral ecosystems and the need for immediate reassessment of global conservation strategies.

The major causes of loss of marine biodiversity include fishing and bycatch; hunting of mammals, birds and turtles; toxic chemicals and nutrient pollution; habitat destruction; and human-assisted transport and release of species to environments where they did not previously exist. Census research is describing and synthesizing globally changing patterns of species abundance, distribution and diversity and modeling the effects of fishing, climate change and other key variables on those patterns. This work is being done across ocean realms and with an emphasis on understanding past changes and predicting future scenarios.

The goal is to synthesize long-term trends and large-scale changes in marine animal populations and ecosystems to add a temporal dimension to spatial patterns and short-term dynamics in marine populations and diversity. Long-term changes being investigated include past trends of depletion and degradation, and also population recoveries. Large-scale changes include shifts in spatial distribution patterns. Investigations of long-term changes and large-scale changes will be accompanied by analyses of the underlying drivers and the consequences of these changes in order to evaluate current and potential future trends in marine biodiversity. Finally, the Census is developing a modeling framework to analyze the consequences of biodiversity changes for the structure and functioning of the food web, today and in the future.

CHAPTER TEN

The Path Forward

The intriguing part about researching marine life is that nearly every time a question is answered, another question is raised.
— IAN POINER, AUSTRALIAN INSTITUTE OF MARINE SCIENCE, CHAIR OF THE CENSUS OF MARINE LIFE SCIENTIFIC STEERING COMMITTEE

To those who follow the news, it comes as no surprise that most ocean ecosystems are in trouble and the ocean's bounty is under siege. Over-fishing, pollution and climate change are taking a toll on the health and vitality of ecosystems throughout the global ocean, as well as on individual species, some of which are perilously close to not surviving the onslaught of human activities. It is becoming increasingly clear that the world's ocean can no longer assimilate all that humans either dump into it or take from it without consequences.

Some news stories of the past decade are shocking, almost unbelievable. They range from the "great Pacific garbage patch" – a flotilla of trash nearly the size of Africa that tends to trap and kill marine life – to reports of massive coral reef bleaching and destruction, to vast dead zones, void of oxygen, that can no longer support life. When you dig deeper, past the headlines, into what scientists have actually discovered, some of the stories become more convincing, and in some cases the news is truly ominous.

Scientists estimate, for example, that half of the coral reefs in the Caribbean and a quarter of the reefs around the world are now dead. It is

Opposite: Critically endangered leatherback turtles (Dermochelys coriacea) routinely cross international borders in the eastern Pacific during extensive migrations over thousands of kilometers. Census research revealed that ocean currents shape the leatherback migration corridor and influence the scope of their dispersal in the South Pacific — results that provide a biological rationale for developing multi-scale conservation strategies along migration corridors and on the high seas.

Sometimes the infrastructure of modern life gets too close to individual human activities. Shown here are divers next to a sewage outfall pipe in Delray Beach, Florida.

believed that this devastation is mainly due to pollution, physical destruction and increasing ocean water temperatures, a direct result of global warming (most coral species can survive only in very narrow temperature ranges). Complicating their survival further is the rise in concentrations of atmospheric carbon dioxide (CO_2). This is causing changes in the ocean's carbonate chemistry system that will affect some of the most fundamental biological and geochemical processes of the sea. Since 1980, ocean uptake of excess CO_2 released into Earth's atmosphere by human activities is significant: about a third has been stored in the oceans. This uptake of CO_2 lowers the ocean's pH, making it more acidic, which in turn causes lower saturation states of the carbonate minerals used to form skeletal structures by many major groups of marine organisms, including corals. Scientists predict that if nothing is done to correct this condition, all of the remaining coral reefs could be dead by 2075.

Another disturbing story is about the large dead zone off the coast of Namibia. Scientists are measuring it as they watch it expand year after year. The only inhabitants that remain in this zone are jellyfish, often described as the cockroaches of the sea in recognition of their ability to

The "great Pacific garbage patch" is a floating mound of garbage in the Pacific Ocean that is nearly the size of Africa. The plastic flotilla is a dangerous hazard for marine life that gets entrapped in its cargo.

thrive under adverse conditions. Not only are jellyfish able to survive in areas with depleted oxygen and rising temperatures, but their populations are exploding as populations of their natural predators – spadefish, sunfish, and loggerhead turtles – decline. According to the U.S. National Science Foundation, jellyfish populations are also thriving and causing problems in Australia, Britain, Hawaii, the Gulf of Mexico, the Black Sea, the Mediterranean, the Sea of Japan and China's Yangtze (Chang Jiang) estuary. The problem is becoming so prevalent that a phrase has been coined to describe it: the jellification of the ocean.

While larger than many, Namibia's dead zone is not unique. A 2008 report in *Science* by Robert J. Diaz, from the Virginia Institute of Marine Science, and Rutger Rosenberg, of the University of Gothenburg in Sweden, suggests that there are now more than 400 dead zones around the world, double what the United Nations reported only two years earlier. Collectively, these dead zones affect a total area in the world's ocean of more than 245,000 square kilometers (98,000 square miles).

Coral bleaching, such as in this reef in the Caribbean, is one result of warming sea temperatures.

Overfishing has been a key player in the decline of marine life populations. According to the Food and Agriculture Organization of the United Nations, an estimated 75 percent of major fisheries are fully exploited, overexploited or depleted. A paper prepared by Census researchers Boris Worm and the late Ransom A. Myers, published in the scientific journal *Nature*, reported that 90 percent of the population of top predators – the likes of tuna, shark and

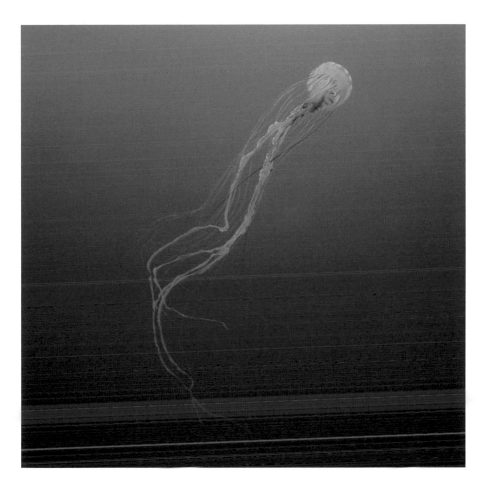

Left: Jellyfish come in all shapes and sizes. Many are quite beautiful, like the relatively common one pictured here, Chrysaora melanaster, *which was photographed in the Arctic's Canada Basin. In some areas of the global ocean, jellyfish are becoming the dominant species.*

swordfish – has already been caught. Without top predators, the entire composition of the food web changes in unanticipated ways, many of which are not necessarily pleasing or healthy. Myers and Worm went on to report in *Science* that if current fishing trends continue, commercial fisheries as we know them will have collapsed by 2050. If this projection proves correct, many of us could witness the end of fish caught in the wild during our lifetimes.

Other scientists suggest that if we continue to fish species to population levels below the numbers required for reproduction, the eventual result could be extinction of those species. The orange roughy is one example often cited. Orange roughy began to be commercially fished only about 10 years ago. It is a long-lived fish, believed to reach roughly 150 years of age when undisturbed, and it begins to reproduce only around the age of 25. A consequence of its growing popularity in kitchens around the world is that this fish is now threatened with extinction. Its breeding adults have been taken out of the population to meet the demand, irrevocably changing population numbers and possibly its future.

Opposite: In this NASA image from space, a flash of blue and green lights up the waters off Namibia in early November 2007, as a phytoplankton bloom grows and fades in the Atlantic Ocean, stretching along hundreds of coastal miles. Phytoplankton blooms – fed by nutrients from both current and upwelling water – are so abundant off Namibia that their death and decomposition often rob the water of dissolved oxygen. As the plants die, they sink to the ocean floor, where bacteria consume them. But there is so much material that the bacteria have used all the oxygen available in the water before they finish breaking down the plants, creating a dead zone where fish can't survive.

In part, the current condition of marine life in the global ocean is a result of exponential advances in technology accompanied by unprecedented growth in the human population over the past century. Ironically, it is the same technological know-how and innovation that has made it possible for explorers to observe the darkest depths of the most remote oceanic areas that has also provided the means to fish expertly and efficiently, catching heretofore unimaginable quantities to feed the world's burgeoning population. In 2008 the world's population was nearing 6.7 billion people; it is projected to reach 9 billion by 2042. This population explosion will create unprecedented demands on global ecosystems and will continue to pose challenges to the health of the global ocean's biological resources.

THE CENSUS HAS MADE A DIFFERENCE

The ominous projections above serve only to reinforce the importance of the Census's work and its potential contribution for the years ahead. There is no question that the Census has already significantly increased understanding of life in the global ocean. Its results hold promise for crafting scientifically based solutions to the problems already outlined, and to others that may subsequently develop.

In 2010 the Census will present the first-ever catalog of marine life, including data on abundance, distribution and diversity, which will serve as a baseline from which future changes in populations can be measured. This in turn will provide policy and decision makers around the globe with a scientific foundation from which to make future management decisions, and possibly to take action to help preserve and restore marine life resources to healthy levels.

The historical aspect of the Census's work demonstrates the advantages of working in multidisciplinary teams to tackle complex scientific questions. Ultimately, its comprehensive approach has broadened the way science is being practiced. The historic baselines for marine species and ecosystems developed by the Census may also prove helpful in managing recovery of threatened resources by setting responsible goals, based on past levels of healthy abundance, distribution and biocomplexity. Biocomplexity is a broad-ranging concept that takes into consideration the interplay of behavioral, biological, chemical, physical and social interactions that affect, sustain or are modified by living organisms, including humans.

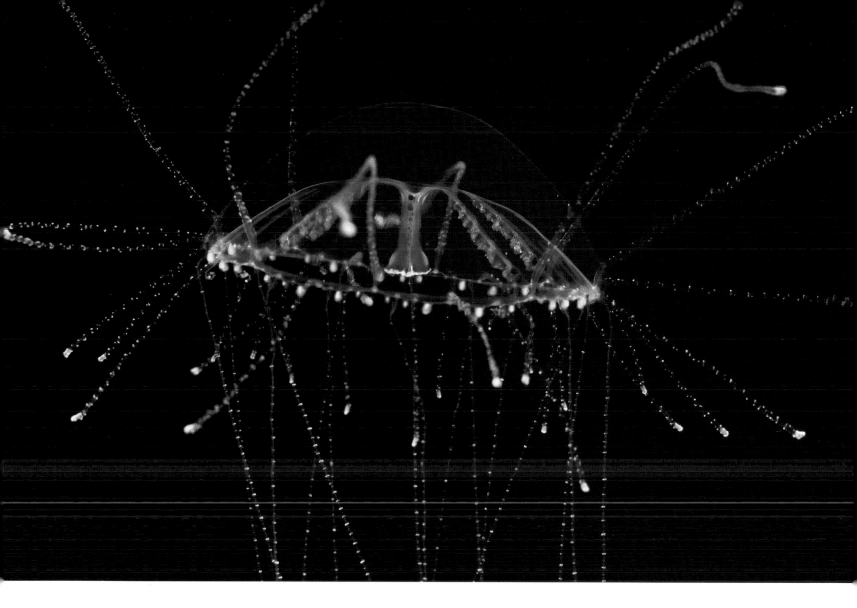

In addition to these valuable management tools, the Census has opened a window into what was once an opaque ocean. The project established an ocean tracking network in the northwest Pacific Ocean that serves as a model for a truly global tracking network; this network will provide not only data about animal behavior but oceanographic data as well. In its expanded global version, thousands of marine animals around the world – from fish to birds to polar bears – will be tracked using acoustic signals. This global network will increase knowledge about where animals travel, while at the same time compiling a comprehensive record of oceanographic information that includes water temperature and salinity. These data will be available to the scientific community at large and will contribute to understanding global climate change. This initiative would not have been possible if Census scientists had not envisioned and taken the first steps toward a system for investigating the world's ocean.

This water jelly (a new species of Olindias*) was photographed during a Census expedition to Lizard Island, part of the Great Barrier Reef in Australia. It is but one of the estimated 230,000 known species that will be cataloged in the online* Encyclopedia of Life.

The Census also has been at the forefront in the use of animals as ocean observers, which makes it possible to experience the watery environment as marine animals do. Over the course of its first eight years, Census scientists tracked 23 different species – from sharks and squid to sea lions and albatrosses – literally "tagging along" as the animals ate, mated, slept and traveled from place to place. This collective view into the lives and behaviors of many creatures had heretofore not been possible. The knowledge gained from this massive tagging initiative is laying the groundwork for scientifically based policies that may serve to protect various species' migration routes and breeding areas to help ensure their future survival.

Census scientists have made significant contributions to what can be seen below the surface through their willingness to test and be the first to implement new underwater equipment and technology. Their innovative approaches include attaching high-definition video cameras to underwater gear to provide real-time, in-depth views of life at great depths; deploying unique trawl configurations that filter large volumes of water to catch tiny drifting zooplankton at depths not sampled before; and being among the first to perform DNA barcoding of plankton at sea. These new technologies have enabled Census scientists to expand the scope of research and improve the efficiency of how it is conducted. Plankton barcoding at sea, for example, accelerated what would have taken three years to accomplish with traditional methods to achieve results in just three weeks. This novel approach to real-time identification is likely to revolutionize the way researchers identify creatures, both on land and at sea.

During its relatively brief existence, the Census has contributed significantly to an understanding of the marine environment and the inhabitants of the global ocean. It has achieved much of what it set out to do, not the least of which is the establishment of an international collaborative network whose participants are willing to share data and pool resources to conduct research on a global scale. Its scientists have participated in more than 100 research expeditions around the globe, from top to bottom and from nearshore to the deep ocean depths. In studying marine life in such places as the coldest locations on the planet, the deepest known active hydrothermal vent and areas recently uncovered by melting Antarctic ice, Census scientists have found life in unexpected places and with a greater diversity and distribution than expected. In terms of abundance they were surprised to find that rare is common.

Opposite: The beauty of this landscape belies the damage caused by terrestrial runoff, a major source of pollution in the global ocean. Finding ways to control runoff is but one of the challenges that must be confronted to ensure future healthy marine-life populations.

While these efforts have increased knowledge of what is known, they have also served to bring scientists closer to comprehending just how much is unknown – and what may never be known – about this vast watery place called the global ocean.

By all accounts, the amount of new knowledge generated by this collective effort has been remarkable, but as Ian Poiner, chief executive officer of the Australian Institute of Marine Science and chair of the Census Scientific Steering Committee, suggests, it is only a beginning, for "each answer poses a new question." And the questions are many and great. The roles of climate change, human activities and future population growth are part of an unpredictable equation in which humans may hold the key to the ultimate outcome of what lives in the ocean. Many of the Census findings are troubling: loss of 90 percent of top predators in the global ocean; the ineffectiveness of marine protected areas in protecting coral reefs; human exploitation as the primary cause of estuarine damage, to name a few.

The Census of Marine Life will produce a digest that summarizes what has been learned about what lived and lives in the world's ocean, but the biggest question of all may remain unanswered. What will live in the ocean of tomorrow may depend largely on what decisions are made to preserve and protect marine life now and into the future. Census scientists are providing the international community with tools to make sound management decisions, but whether that community chooses to do so is virtually anyone's guess. Perhaps in the years ahead the answer will become clearer as scientists continue to search, lured by the magic beauty and mystery of life below the water's surface.

Census expeditions were undertaken from pole to pole, from nearshore to open ocean, from at the surface to miles below. Here a Census researcher studies gentoo penguins (Pygoscelis papua) *encountered during a seal-tagging expedition to Antarctica.*

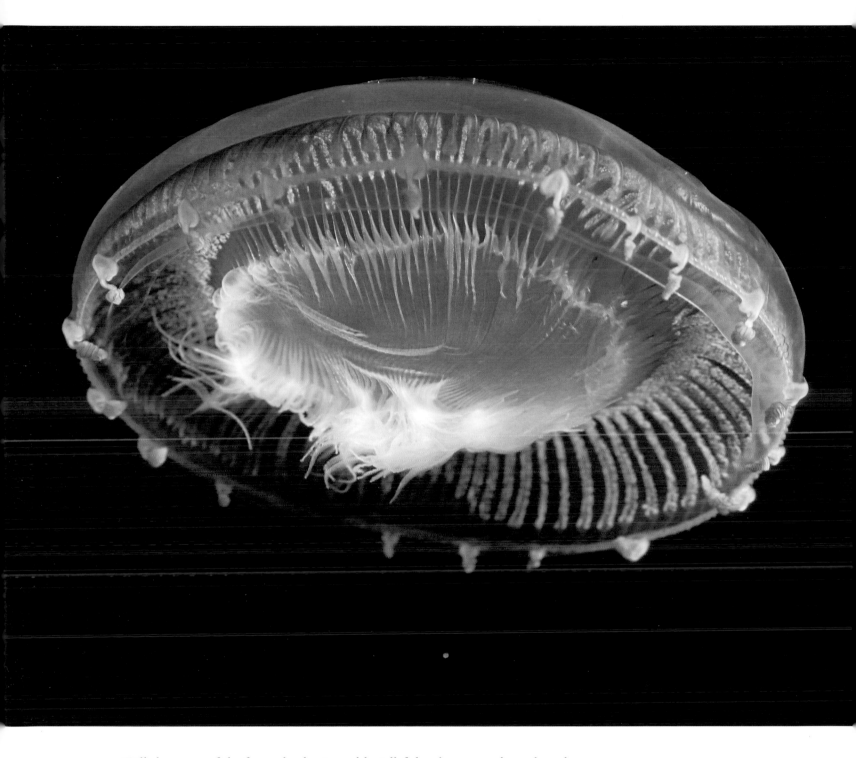

Will the ocean of the future be dominated by jellyfish – known as the cockroaches of the sea because of their ability to thrive under adverse conditions? Shown here is a beautiful example of this adaptable species, Aequorea macrodactyla, *photographed during a Census expedition to the Celebes Sea.*

GLOSSARY

Abundance The amount of individuals in a given species.

Abyssal plains Large, relatively flat regions on the deep sea floor from approximately 4,000 to 6,000 meters (2.5–3.75 miles) in water depth.

Acoustic tag An electronic tag attached to an animal that uses a sound signal to transmit information about the tagged animal's location, including data on its depth, the surrounding water temperature and amount of light in the water.

ALV (autonomous lander vehicle) A metal frame placed on the ocean floor that is outfitted with photographic equipment to record time-lapse photographic images of marine life on the ocean floor, and that can be equipped with instruments to measure physical properties of the ocean such as conductivity, temperature, depth and current speed.

Amphipods Refers to the order of Amphipoda that includes more than 7,000 described species of shrimp-like crustaceans ranging in size from 1 mm to 140 mm in length.

Anaerobic An environment with an absence of oxygen.

Anemone (short for sea anemone) A sedentary marine invertebrate with a columnar body that bears a ring of stinging tentacles around the mouth.

Animal-borne instrumentation Packages of sensors, carried by animal hosts, which detect and collect data about various biological or environmental parameters.

Archival acoustic tag A new generation of acoustic tag which contains sensors that detect and store depth and water temperature data. As the animal bearing the tag swims past a fixed acoustic receiver, the tag relays this environmental data along with the unique identification code of that tag/animal.

Argos satellite telemetry A satellite-relaying technology that allows researchers to track animal movement and environmental conditions worldwide.

ARMS (autonomous reef monitoring structure) An artificial habitat— "underwater dollhouse"—made of PVC plastic and mimicking the nooks and crannies of a natural reef, which is deployed into the reef environment and monitored over time, allowing scientists to study the re-colonization of reef space.

Arthropods Refers to Arthropoda, a large phylum of invertebrate animals that includes insects, spiders, crustaceans and their relatives. Approximately 80 percent of all known animal species belong to Arthropoda. Arthropods have a segmented body, an external skeleton and jointed limbs, and are sometimes divided among taxonomic classes.

Ascidians Sea squirts.

AUV (autonomous underwater vehicle) An unmanned exploration vehicle that is not tethered to a support ship.

Baseline A reference point used to characterize an ecosystem and from which future change can be determined and quantified.

Bathymetric Refers to bathymetry, the study of water depth in oceans, seas or lakes.

Benthic Refers to the region of the seafloor, including the flora and fauna and bottom sediments of a sea, lake or other body of water.

Benthic grab Equipment used to sample the seafloor by literally "taking a bite" out of the seafloor and bringing it back to the surface for study.

Biocomplexity A broad-ranging concept that takes into consideration the interplay of behavioral, biological, chemical, physical and social interactions that affect, sustain or are modified by living organisms, including humans.

Biodiversity The variety of life on Earth or in a particular habitat or ecosystem.

Biologging The practice of logging and relaying physical and biological data using tags attached to animals.

Biological hotspot A concentrated area of high productivity and, usually, high biodiversity in the ocean.

Biomass The total mass of living biological organisms in a given area or ecosystem at a given time.

Biota The animal and plant life of a particular region, habitat, or geological period.

Bivalves An aquatic mollusk that has a compressed body enclosed within two hinged shells, such as oysters, clams, mussels and scallops.

Brine Water saturated or nearly saturated with salt.

Bristle worm A marine annelid worm that has a segmented body with numerous bristles on the fleshy lobes of each segment. Also called a polychaete.

Bryozoans Tiny colonial animals that generally build stony skeletons of calcium carbonate. They are superficially similar to coral.

Bycatch Fish caught by fishers fishing for other species.

Calcium carbonate A chemical compound used by many marine creatures to construct their shells and skeletons.

Calved Refers to when a large piece of ice has split off from an iceberg or broken off the face of a glacier.

Cephalopod Refers to Cephalopoda a class of active predatory mollusks comprising octopuses, squids, cuttlefish and the "living fossil," nautilus. They have a distinct head with large eyes and a ring of tentacles around a beaked mouth, and many are able to release a cloud of inky fluid to confuse predators.

Cheliped One of the pair of legs that bears the large chelae (claws) in crustaceans.

Chemosynthetic ecosystems Areas where life depends on chemical processes rather than on photosynthesis.

Cnidarians Refers to Cnidaria, a phylum of some 9,000 aquatic animal species, mainly marine. Their distinguishing feature is cnidocytes, specialized cells used for catching prey.

Coelenterate An obsolete term that refers to two animal phyla, Ctenophora (comb jellies) and Cnidaria (coral animals, true jellies, sea anemones, sea pens and others). They have only two layers of cells, external and internal.

Cold seeps Areas of the ocean floor where hydrogen sulfide, methane and other hydrocarbon-rich fluids, the same temperature as the surrounding seawater, slowly seep from the seafloor.

Colonial Of, relating to, or characteristic of a colony or colonies.

Continental margins The submerged continental shelf and slope that forms the outer edge of a major land mass.

Continental rise The area below the continental slope that merges into the deep ocean floor or abyssal plain.

Continental shelf A very gradually sloping border between the edge of a continent and the ocean basin.

Copepods Small aquatic crustaceans, many of which occur in plankton and some of which are parasitic on larger aquatic animals.

Cretaceous/Teritiary (K-T) boundary Approximately 65 million years ago, when the dinosaurs and ammonite cephalopods died out.

Crinoids Ancient marine animals that make up the class Crinoidea of the echinoderms. They live both in shallow water and in depths as great as 6,000 meters (20,000 feet). Also known as sea lilies or feather stars.

Cross-boundary A condition affecting resource management efforts, where the resource in question is mobile and may move across political boundaries, thus being able to be targeted by more than one adjacent user group.

Crustacean A large group of mainly aquatic invertebrates that include crabs, lobsters, shrimps, wood lice, barnacles and many minute forms. Most have four or more pairs of limbs and several other appendages.

CTD Tag (conductivity/temperature/depth tag) a new generation of satellite-relayed data logger developed by the Sea Mammal Research Unit at Scotland's University of St. Andrews. This is the standard technology used for many modern tagging applications such as seal and leatherback turtle research.

Ctenophores Refers to Ctenophora, a small phylum of aquatic invertebrates that comprises the comb jellies.

Decapods "Ten-footed"; refers to an order of crustaceans that includes many familiar species such as lobsters, crabs, crayfish, prawns and shrimp. Most decapods are scavengers. Molluscan squid also have 10 appendages and share the name Decapoda.

Diatoms Single-celled algae.

Distribution Where species live or in some cases higher taxa.

Diversity Number of species or in some cases higher taxa.

DNA barcoding An identification approach that uses a small segment of an organism's DNA to identify its species and treats this DNA sequence like a barcode for future identification.

DTV (deep-towed vehicle) An unmanned vessel pulled behind a research vessel as it traverses the ocean. It is used to measure biological, physical and chemical aspects of the ocean.

Echinoderms Refers to Echinodermata, a phylum of marine invertebrates that includes starfishes, sea urchins, brittlestars, crinoids and sea cucumbers. They have fivefold radial symmetry, a calcareous skeleton, and tube feet operated by fluid pressure.

Echolocation The location of objects by using reflected sound; in particular it is used by animals such as dolphins and bats.

Ecosystem-based management A system of resource management that approaches the protection, conservation and use of living resources from the point of view of maintaining balance in the entire ecosystem, rather than targeting management efforts at single or small groups of species.

Elasmobranchs The subclass of cartilaginous fish that includes skates, rays and sharks.

Endemic Refers to organisms that are exclusively native to a place or plant and animal life of a region (biota).

Epifauna Marine animals that live on the seafloor.

Estuary The tidal mouth of a large river where it meets the sea.

Extinction event A sharp decrease in the number of species in a relatively short period.

Fauna The animals of a particular region, habitat or geological period.

Food web A system of interlocking and interdependent food chains or complex interactions between plants and animals that tells us who is eating what or whom.

Gadids Refers to a family of generally medium-sized marine fish, Gadidae, which includes cod, haddock, whiting and pollock. Most species of gadids are found in the temperate waters of the Northern Hemisphere, although there are some exceptions.

Gas hydrates Water-based crystalline solids that resemble ice crystals with gas trapped inside.

Gastropod Refers to Gastropoda, a large class of mollusks, which includes snails, slugs, whelks and all terrestrial kinds. They have a large muscular foot for movement and (in many kinds) a single asymmetrical spiral shell.

Gelatinous Having a jelly-like consistency.

GIS (geographic information system) A computer mapping technology that visually represents measurements of many different types of physical and biological characteristics for a specific geographical area.

Global conveyor belt A name used for the thermohaline circulation of the global ocean.

Glypheid Refers to Glypheoidea, a group of lobster-like decapod crustaceans that form an important part of fossil faunas.

Habitat A particular type of environment regarded as a home for organisms.

Highly migratory A condition characterizing certain animal species where they regularly move across large areas and may, therefore, cross political boundaries and be subject to use pressure from more than one user group.

Hydrates A compound, typically a crystalline one, in which water molecules are chemically bound to another compound or an element.

Hydrothermal vents Places on the ocean floor where a continuous flow of superheated, mineral-rich water flows up through cracks in Earth's crust.

In situ Situated in an original, natural or existing place.

Invertebrate An animal lacking a backbone.

Island arc A curved chain of volcanic islands formed by the movements of Earth's crustal plates.

Isopods Crustaceans that include the terrestrial wood lice and several marine and freshwater parasites. They have a flattened segmented body with seven similar pairs of legs.

Isotopes Naturally occurring forms of an element that have different numbers of neutrons in their atomic nuclei.

Krill A small, shrimp-like planktonic crustacean of the open seas. It is eaten by a number of larger animals such as whales, penguins and seals.

Listening curtain A linear array of acoustic receivers that form a boundary. Communication between the array and acoustic tags carried by animal hosts allow scientists to track animal movements across the boundary and through a range.

Longlining Among the most widespread of fishing methods in which miles of line with baited hooks are used to catch a wide range of fish species.

Magnanese nodule A small concretion (a hard solid mass formed by the local accumulation of matter) consisting of manganese and iron oxides, occurring in large numbers on the ocean floor, usually in very deep water.

Marine microbes Microscopic marine life.

Massif A compact group of mountains, especially one that is separate from other groups.

Megalops Crab larval stage.

MOCNESS (multiple opening/closing net and environmental sampling system) Trawl nets fabricated from very fine nylon mesh that were used to collect zooplankton in deep water from multiple depths.

Mollusk An invertebrate of a large phylum that includes snails, slugs, mussels and octopuses. Mollusks have a soft, unsegmented body and live in aquatic or damp habitats, and most kinds have an external calcareous shell.

Morphology The branch of biology that deals with the form of living organisms, and with relationships between their structures.

MPAs (marine protected areas) A commonly used term that generally describes an ocean region and its associated flora, fauna, and historical and cultural features that has been reserved by law or other effective means to protect part or all of its environment.

Multibeam sonar Another term for multi-frequency echo sounder.

Multi-frequency echo sounder A sophisticated sonar system used to estimate the size of plankton and fish populations and for species identification.

MVP (moving vessel profiler) A towed vehicle that moves up and down, used to house a video plankton counter and other sampling devices.

Nearshore environment A narrow region of the coastal ocean closely associated with adjacent landmasses and, therefore, directly affected by the people who occupy the adjoining land. Examples include shorelines of all kinds, coastal bays and estuaries, and coral reefs.

Non-colonial Not part of a colony.

OBIS (Ocean Biogeographic Information System) The Census of Marine Life's online interactive database.

Ocean acidification The name given to the ongoing decrease in the pH of the global ocean caused by uptake of increased amounts of carbon dioxide from the atmosphere.

Ocean basin A natural depression on Earth's surface containing salt water.

Oligochaete A type of marine worm.

Ostracod A class of minute aquatic crustaceans that have a reduced number of appendages and a hinged shell from which the antennae and a number of appendages protrude.

Overfishing When too many fish of a species are fished so that stable population levels cannot be sustained.

Oxidation The process or result of oxidizing or being oxidized, which is to combine or become combined chemically with oxygen.

Paleoecology The ecology of fossil animals and plants.

Paleooceanographic Older or ancient, as related to the branch of science that deals with the physical and biological properties and phenomena of the sea.

Paleontological Of or relating to the study of the forms of life existing in prehistoric or geologic times, as represented by the fossils of plants and animals.

Paleozoologist One who studies the ancient behavior, structure, physiology, classification, and distribution of animals from fossils.

Pelagic Free-swimming.

Pelagic environment The open ocean or "blue water" zone.

Permian/Triassic boundary 251 million years ago during which the largest extinction event occurred.

Pinnipeds An order of carnivorous aquatic mammals that comprises the seals, sea lions and walrus, distinguished by their flipper-like limbs.

Phylogenetic The evolutionary development and diversification of a species or group of organisms, or of a particular feature of an organism.

Phytoplankton Microscopic plants drifting or floating in the sea or fresh water.

Plankton The small and microscopic organisms drifting or floating in the sea or fresh water, consisting chiefly of diatoms (single-celled algae), protozoans (single-celled microorganisms), small crustaceans and the eggs and larval stages of larger animals.

Polychaete a type of marine worm.

Pop-up archival tag A data logging tag that is attached to an animal for a fixed period of time before releasing and floating to the surface of the water, where its collected data can be recovered by various means.

Predator An animal that naturally preys on others.

Prey An animal that is hunted and killed by another for food.

Protozoan A single-celled microorganism with a nucleus.

PSAT (pop-up satellite archival tag) A tag that collects the same data as a pop-up archival tag but can transmit data to an orbiting satellite, which relays the information to researchers.

Pseudochelae Pincer-like appendages.

Reef recolonization The rebound or regrowth of reef structure and reef animal communities following disturbances such as pollution events, storm damage or ship groundings, which damage or clear an area of reef space.

Rhodolith A group of marine red algae, which are characterized by hard calcium carbonate (calcareous/chalky) skeletons around which the living algal tissue grows. Rhodoliths form vast beds, creating habitat for various associated species.

Rift Where two plates of Earth's crust are being pulled apart.

ROV A remotely operated vehicle controlled via a cable.

Run Time The length of time that a biologging tag can continue to track and record data.

RV A research vessel.

Satellite-relayed data logger Sophisticated, self-contained units, which are embedded with various sensors that are attached to animals to collect data such as depth, temperature, salinity and swim speed for various periods of time (from days to years).

Satellite remote sensing A technology using reflected light or radar to determine different aspects of ocean conditions such as water temperature; chlorophyll levels, which indicate phytoplankton abundance; and oceanic currents.

Seamounts Typically steep-sided extinct volcanoes that lie below the ocean surface. An "official" seamount must be at least 1,000 meters (3,300 feet) high.

Shell midden A heap of discarded shells.

Side-scan sonar A type of acoustic technology that is used to map the ocean floor and track schools of fish. Pulses of sound are projected by a ship or a towed device. Sound waves that bounce off objects in the water are reflected back to the ship, where instruments convert them into images.

Slurp guns Suction samplers used to collect organisms.

Sp. Unidentified or unspecified species.

Speciation The formation of new and distinct species in the course of evolution.

SPOT (smart position and temperature tag) A family of biologging tags that record an animal's position and the water temperature, as well as its speed and the ambient water pressure (used to indicate depth). This tag is designed to broadcast its data when the animal is at or very near the water's surface.

Standardized investigation protocol A scientific method or procedure that is agreed upon and adopted by a group of investigators in order to ensure quality control, valid comparison and repeatability of results.

Submarine canyon A steep-sided valley on the seafloor of the continental slope.

Submarine trench A trench on the seafloor of the continental slope.

Submersible A type of underwater vessel with limited mobility, which is typically transported to its area of operation by a surface vessel. Submersibles can be either manned or unmanned.

Supercontinent Each of several large landmasses (notably Pangaea, Gondwana and Laurasia) thought to have divided in the geological past to form the present continents.

Suspension feeders Animals that rely on water to deliver food and oxygen.

Symbiont Either of two organisms that live in symbiosis with one another.

Symbiosis Interaction between two different organisms living in close physical association, typically to the advantage of both.

Taxonomic analysis A type of biological analysis where genetic and morphological characteristics of species are used to determine the relationships between and among organisms so that scientists can appropriately name them and organize their evolutionary ties.

Thermohaline circulation Ocean circulation that is driven by differences in water-mass densities rather than being driven by wind.

Time/depth recorders A family of tag used in biologging that logs the dive time and dive depth of marine mammals such as elephant seals.

Transboundary dispute A condition in which a species occupies a range, or migrates through a range, that includes more than one political unit and is sought after or used by adjacent user groups, creating conflict over issues of management and equitable use.

Trawls Specialized nets used to collect marine specimens; usually wide-mouthed nets that are dragged through the water behind a vessel.

Trophic history An organism's historical position in the food chain. (Some people use this in the sense that you can determine an animal's "trophic history," i.e., what it has been eating, by analyzing the composition of its fat or isotope ratios.)

Trophic level Each of several hierarchical levels in an ecosystem, comprising organisms that share the same function in the food chain and the same nutritional relationship to the primary sources of energy.

Tube worm A marine bristle worm, that lives in a tube made from sand particles or in a calcareous tube that it makes by secreting calcium carbonate from sea water.

Upwelling An oceanographic phenomenon that occurs when winds drive water away from a region of the ocean surface, allowing deeper waters to rise up to fill the gap.

Vampyromorph Refers to Vampyromorphida, an order of cephalopods.

VPR (video plankton recorder) A towed box in which water moves past a video camera that records images of plankton either continuously or at predetermined times.

Xenophyophores Marine protozoans; giant single-celled organisms found throughout the world's ocean, but in their greatest numbers on the abyssal plains of the deep ocean.

Zooplankton Plankton consisting of small animals and the immature stages of larger animals.

FURTHER READING

A complete searchable list of Census of Marine Life publications is available at http://db.coml.org/comlrefbase/. Other books of potential interest include the following titles.

Antczak, Andrze and Roberto Cipriano, eds. *Early Human Impact on Megamolluscs.* London: Archaeopress, Publisher of *British Archaeological Reports,* 2008.

Baker, Maria, Brigitte Ebbe, Jo Hoyer, Lenaick Menot, Bhavani Narayanaswamy, Eva Ramirez-Llodra and Morten Steffensen. *Deeper Than Light.* Bergen, Norway: Bergen Museum Press, 2007.

Carson, Rachel. *The Edge of the Sea.* Boston: Mariner Books, 1998

Census of Marine Life on Seamounts Data Analysis Working Group. *Seamounts, Deep-Sea Corals and Fisheries: Vulnerability of Deep-Sea Corals to Fishing on Seamounts beyond Areas of National Jurisdiction.* Cambridge: UNEP World Conservation Monitoring Centre, 2006.

Clover, Charles. *The End of the Line: How Overfishing Is Changing the World and What We Eat.* Berkeley: University of California Press, 2008.

Corson, Trevor. *The Secret Life of Lobsters: How Fishermen and Scientists Are Unraveling the Mysteries of Our Favorite Crustacean.* New York: Harper Perennial, 2005.

Cousteau, Jacques. *The Human, the Orchid, and the Octopus: Exploring and Conserving Our Natural World.* New York: Bloomsbury, 2008 (reprint edition).

Earle, Sylvia. *Sea Change: A Message of the Oceans.* New York: Ballantine Books, 1996.

Earle, Sylvia and Linda Glover. *Ocean: An Illustrated Atlas.* Washington, D.C.: National Geographic Society, 2008.

Ellis, Richard. *Singing Whales and Flying Squid: The Discovery of Marine Life.* Guilford, CT: Globe Pequot Press, 2006.

Ellis, Richard. *Tuna: A Love Story.* New York: Knopf, 2008.

Field, John G., Gotthilf Hempel, and Colin P. Summerhayes, eds. *Oceans 2020: Science, Trends, and the Challenge of Sustainability.* Washington, D.C.: Island Press, 2002.

Grescoe, Taras. *Bottom Feeder: How to Eat Ethically in a World of Vanishing Seafood.* New York: Bloomsbury, 2008.

Koslow, Tony. *The Silent Deep: The Discovery, Ecology, and Conservation of the Deep Sea.* Chicago: University of Chicago Press, 2007.

Monsalve, Héctor Elias and Pablo Enrique Penchaszadeh. *Patagonia Submarina / Underwater Patagonia.* Buenos Aires: Ediciones Larivière, 2007.

Nouvian, Claire. *The Deep: The Extraordinary Creatures of the Abyss.* Chicago: University of Chicago Press, 2007.

Pitcher, Tony J., Paul J.B. Hart, Telmo Morato, Malcolm R. Clark, Nigel Haggan and Ricardo S. Santos, eds. *Seamounts: Ecology, Fisheries and Conservation.* Oxford: Wiley Blackwell, 2007.

Prager, Ellen. *Chasing Science at Sea: Racing Hurricanes, Stalking Sharks, and Living Undersea with Ocean Experts.* Chicago: University of Chicago Press, 2008.

Roberts, Callum. *The Unnatural History of the Sea.* Washington, D.C.: Island Press, 2007.

Sloan, Stephen. *Ocean Bankruptcy: World Fisheries on the Brink of Disaster.* Guilford, CT: The Lyons Press, 2003.

Starkey, David J., Poul Holm and Michaela Barnard. *Oceans Past: Management Insights from the History of Marine Animal Populations.* London: Earthscan, 2008.

Thorne-Miller, Boyce and Sylvia Earle. *The Living Ocean: Understanding and Protecting Marine Biodiversity.* Washington, D.C.: Island Press, 1999.

Wehrtmann, Ingo S. and Jorge Cortés, eds. *Marine Biodiversity of Costa Rica, Central America.* New York: Springer, 2008.

Wilkinson, C., ed. *Status of Coral Reefs of the World: 2008.* Townsville, Australia: Global Coral Reef Monitoring Network, 2008.

PHOTO CREDITS

114	Photograph by Margaret Butschler, courtesy of Vancouver Aquarium
115	Courtesy Josh Adams
116	Courtesy G. Chapelle / AWI
118	Courtesy Elizabeth Calvert Siddon, NOAA
119	Top and Bottom: Courtesy NASA
120	Courtesy Bodil Bluhm
121	Image courtesy of The Hidden Ocean, Arctic 2005 Exploration, http://oceanexplorer.noaa.gov/explorations/05arctic/welcome.html
122	Top left: Courtesy Kevin Raskoff
	Top right: Courtesy Bodil Bluhm / Katrin Iken
	Bottom left: Courtesy Bodil Bluhm
	Bottom right: Courtesy Bodil Bluhm / Katrin Iken
123	Top: Courtesy Bodil Bluhm
	Bottom: Image courtesy of The Hidden Ocean, Arctic 2005 Exploration, http://oceanexplorer.noaa.gov/explorations/05arctic/welcome.html
124	Courtesy George Walker
125	Courtesy G. Chapelle / AWI
126	Courtesy G. Chapelle / AWI
127–128	Courtesy J. Gutt, © AWI / MARUM, University of Bremen
130	Courtesy John Mitchell IPY-CAML
131	Courtesy J. Gutt, © AWI / MARUM, University of Bremen
132	Image courtesy of The Hidden Ocean, Arctic 2005 Exploration, http://oceanexplorer.noaa.gov/explorations/05arctic/welcome.html
133	Courtesy Map Resources
134	Top: Image courtesy of The Hidden Ocean, Arctic 2005 Exploration, http://oceanexplorer.noaa.gov/explorations/05arctic/welcome.html
	Bottom: Courtesy Dan Costa
135	Map courtesy of the Australian Antarctic Division © Commonwealth of Australia 2007 SCAR Map Catalogue No. 13353
136	Courtesy Andrzej Antczak
138	Courtesy Susan Middleton
139	Courtesy Jim Maragos
140	Courtesy Cory Pittman
141	Courtesy NOAA / PNMN
142	Courtesy National History Museum of LA County
143	Courtesy Peter Lawton
144	Courtesy Robin Rigby
145	Courtesy Brenda Konar
146	Top: Courtesy Katrin Iken / Casey Debenham
	Bottom: Courtesy Cory Pittman
147	Photo by Stefano Schiaparelli, courtesy of Santo 2006 Global Biodiversity Survey
148	Courtesy National History Museum of LA County
149–150	Courtesy Gustav Paulay
151	© Donald Miralle / Getty Images
152	Courtesy Emory Kristof / National Geographic Image Collection
154	Pacific Ring of Fire 2004 Expedition. NOAA Office of Ocean Exploration; Dr. Bob Embley, NOAA PMEL, Chief Scientist
155	Courtesy Ricardo Santos
156	Courtesy NOAA
157	Courtesy S. Hourdez and C.R. Fisher
158	Courtesy MARUM, University of Bremen
159	Top: Courtesy C. Fisher / L. Levin / D. Bergquist / I. MacDonald
	Bottom: Courtesy Ian MacDonald
160	Top: Courtesy NOAA / NIWA
	Bottom: Courtesy Mountains in the Sea, 2004. NOAA Office of Ocean Exploration; Dr. Les Watling, Chief Scientist, University of Maine
161	Courtesy NURC / UNCW and NOAA / FGBNMS
162	Courtesy NOAA, http://oceanexplorer.noaa.gov/explorations/03mexbio/feb13/media/feb13pic.html
163	© NORFANZ / Te Papa, MA°I.137219. Montage by Rick Webber, photography by Karin Gowlett-Holmes. This image has been provided courtesy of the NORFANZ partners – Australia's Department of the Environment, Water, Heritage and the Arts and CSIRO and New Zealand's Ministry of Fisheries and NIWA. For more information on the voyage, visit http://www.environment.gov.au/coasts/discovery/voyages/norfanz/index.html. The use of this image does not imply the endorsement of the NORFANZ partners of the content of this article.
164–165	Courtesy Mountains in the Sea Research Team/Institute for Exploration / NOAA
167	Top: Courtesy Les Watling / NOAA
	Bottom: Gulf of Alaska 2004. NOAA Office of Ocean Exploration
168	Courtesy NOAA / Monterey Bay Aquarium Research Institute
169–172	Courtesy Brigitte Ebbe
173	Gary Cranitch, Queensland Museum
174	Courtesy Cedric D'Udekem D'Acoz, Royal Belgian Institute of Natural Sciences
176	Courtesy Wiebke Brokeland
177	Top: Courtesy Wiebke Brokeland
	Bottom: Courtesy G. Chapelle / AWI
178	Courtesy J. Groeneveld
179	Courtesy Slovin Zanki
181	Top: Courtesy Janet Bradford-Grieve, NIWA
	Bottom: Courtesy Dhugal Lindsay
182	Courtesy B.Richer de Forges / J.Lai-IRD
183	Courtesy Marie-Catherine Boisselier
184	Courtesy Russ Hopcroft
186	Courtesy Richard Pyle
187	Left: Courtesy Kevin Raskoff
	Right: Courtesy Bodil Bluhm
188	Top: Courtesy R. Gradinger and B. Bluhm / UAF / ArcOD
	Bottom: Courtesy Kevin Raskoff
189	Top: Courtesy Russ Hopcroft
	Bottom: Courtesy Wiebke Brokeland
190	Top: Courtesy Dorte Janussen, Senckenberg, FRG
	Bottom: Images courtesy of the National Snow and Ice Data Center
191–192	Courtesy Cedric D'Udekem D'Acoz, Royal Belgian Institute of Natural Sciences
193	Courtesy G. Chapelle / AWI
194	Top: Courtesy Leanne Birden
	Bottom left and right:
	Courtesy Steven Haddock
195	Left: Courtesy MAR-ECO, Peter Rask Möller, 2006
	Right: Courtesy Tomio Iwamoto, California Academy of Sciences, 2004
196	Top: Courtesy IFREMER, A. Fifis, 2006
	Bottom: Courtesy Yunn-Chih Liao / Kwang-Tsao Shao, Academia Sinica, 2007
197	Top: Courtesy Jan Michels / COMARGE
	Bottom: Courtesy Yousra Soliman, Texas A&M University
198	Top: Courtesy Jed Fuhrman, University of Southern California
	Bottom: Courtesy David Patterson
199	All: Courtesy David Patterson
200	Carola Espinoza
202–203	Courtesy David Shale
204–207	© www.davidfleetham.com
208	Courtesy Jo Høyer
209	Courtesy NOAA
210	Top: Courtesy Commander John Bortniak, NOAA Corps (ret.)
	Bottom: © www.davidfleetham.com
211	Top: Courtesy Michael Penn
	Bottom: © www.davidfleetham.com
213	© Shedd Aquarium / photo by Brenna Hernandez
214	Provided by racerocks.com, Lester B. Pearson College, Victoria, BC
215	All: © www.davidfleetham.com
216	Top and middle: © www.davidfleetham.com
	Bottom: © Kanagawa Prefectural Museum of Natural History (Photo: T. Suzuki) catalog number: KPM-N0054758
217	© www.davidfleetham.com
219	Courtesy Captain Philip A. Sacks
220	Courtesy Florida Fish and Wildlife Conservation Commission / NOAA
222	© Jason Bradley / BradleyPhotographic.com
224	Courtesy Steve Spring Reef Rescue
225	Top: Courtesy Clarita Natoli
	Bottom: Courtesy Tyler Smith
226	Courtesy NASA
227	Courtesy Katrin Iken
229	Courtesy Gary Cranitch, Queensland Museum
231	Copyright Mark DeFeo 2008. Heliphotos@mchsi.com All rights reserved
232	Courtesy Dan Costa

INDEX